Managing for Long-term Community Recovery in the Aftermath of Disaster

Daniel J. Alesch, Lucy A. Arendt, and James N. Holly

Public Entity Risk Institute

Public Entity Risk Institute is a tax exempt nonprofit whose mission is to serve public, private, and nonprofit organizations as a resource to enhance the practice of enterprise risk management. For more information on PERI, visit the organization's website: www.riskinstitute.org

Public Entity Risk Institute
11350 Random Hills Road, Suite 210
Fairfax, VA 22030
(p) 703.351.1846
(f) 703.352.6339
www.riskinstitute.org

ISBN 978-0-9793722-2-3

Volume 1, 2008

Printed in the United States of America

CONTENTS

TABLES AND FIGURES

Tables

Figures

ACKNOWLEDGMENTS

OVER THE PAST FIFTEEN YEARS we have talked with and listened to hundreds of public officials, small business owners, hospital administrators, bankers, professionals, executives, and others who, with their families, lived through one or more disaster events, the immediate response, and the short- and long-term aftermath. They generously took time to talk with us about their experiences following one or more extreme events. They spoke frankly and from the heart. The memories we evoked often brought tears. These people helped us understand what happened to their businesses, their communities, their families, and themselves.

We visited many of the disaster sites three or more times over the years following the event and were graciously welcomed with each visit. We gratefully acknowledge their generous contributions of time, experience, and insights. We respect the privacy of all those with whom we spoke in confidence and have taken pains to maintain their anonymity. We do, however, want to explicitly express our gratitude to Dorothy Zaharako for all her help. She has experienced more than her share of hurricanes and is currently the Financial Services Director of Stuart, Florida. We also appreciate all the insights Sheridan "Butch" Truesdale has shared with us from his vantage point in Florida.

We also thank our professional colleagues who contributed opinions, knowledge insights, and reviews of this project. Their professional efforts and dedication to reducing the risk and consequences of natural disasters is inspiring. We particularly want to thank William Anderson, Peter May, William Petak, and Susan Tubbesing for their thoughtful comments on earlier versions of this document and Sarah Nathe who did a splendid job, as always, of editing that earlier version. We extend thanks to Dennis Wenger who inspired us to think deeply into recovery processes. We extend our appreciation to those who helped us in the early years in our study of consequences and recovery, notably Elliott Mittler, A. Sam Ghanty, Robert Nagy, and Michael Meulemans.

Without the Public Entity Risk Institute this and similar projects would have never seen the light of day. We are greatly indebted to PERI for its support and patience over the years and extend our sincere thanks to PERI's exceptional staff.

We extend a special thanks to Gerry Hoetmer, Claire Reiss, and Jessica Hubbard for their assistance in completing this project.

We also express appreciation to Jane Cotnoir for her exceptional editing and good humor as well as a sincere thanks to Colleen Gratzer who designed the proof pages and the cover — thank you for your excellent work.

All that having been said, we, of course, are solely responsible for what we have written.

Daniel J. Alesch, Ph. D.
Lucy A. Arendt, Ph. D.
James N. Holly, Ph. D.
Green Bay, Wisconsin

CHAPTER 1

AFTER THE DISASTER: THE CHALLENGE OF RECOVERY

"Bad things happen when you least expect them."
Raymond Chandler's Phillip Marlowe

*T*HIS BOOK HAS BEEN PREPARED with two primary audiences in mind. The first is local officials and community leaders in the United States in an effort to help them understand the less obvious, but critically important, consequences of extreme events, and to provide guidance on what to do to help ensure long-term community recovery. A second audience is public officials who devise, implement, and evaluate policies and programs intended to facilitate local recovery. We hope that others who want to understand disaster consequences and recovery processes will find this book useful as well.

The terms that are commonly used in connection with community disaster are, in our estimation, far too simple and nondescriptive. People generally know what emergency response is: responding to the emergencies associated with extreme events. Then, too often, they refer to restoring basic services and rebuilding the physical community as "recovery." In general, we are very good at restoring basic services, and we are often good at rebuilding the physical artifacts of the community. We are much less skilled, however, at what is sometimes called "long-term community recovery," which consists of restoring or building anew the social, political, and economic elements of the community fabric that make a community viable over the long haul. This book focuses on that last part: what we call "community recovery."

Managing for Long-Term Community Recovery in the Aftermath of Disaster is based on more than a decade of field research in more than two dozen U.S. communities that have experienced disasters. In discussions with hundreds of public officials, business owners, thoughtful observers, and ordinary citizens, we gained some understanding that we think will prove useful to local officials, civic leaders, and officials at higher levels of government. For example, we found that many local government officials and civic leaders face three major obstacles when working on recovery. First, they typically know what kinds of physical damage to expect in the built environment, but only rarely do they understand the full extent of the consequences that cascade into the social and economic life of the community. Second, few officials at any level of government have much understanding of community recovery processes and of how to help them along. Third, most local officials have relatively little knowledge about what happened in other communities in the months and years following a disaster. This book is intended to help overcome those three obstacles.

Throughout the course of our research, we asked lots of questions and were asked lots of questions in return. Local officials, who already knew how to get the utilities up and running, remove debris, and rebuild infrastructure, were faced with local economies that were in shambles, people moving out and others moving in, housing shortages, and a host of other problems. Seeking help they frequently asked, almost rhetorically, "Who do I call?" Although we can't provide a phone number, this is our attempt to answer that question.

This book does not address what to do in the first few days or weeks after an extreme event, the period when volunteers and professional crews are removing debris, when workers are struggling to restore public utilities, when the Red Cross and others are providing food and temporary shelter for those displaced, and when FEMA is there to render aid. Those are critical times, but local government officials and emergency managers generally know what to do during that period and how to do it, and if they do not, a significant amount of information is available to guide them. Instead, this book is about the period after the mud dries and most of the volunteers have gone home— the long aftermath during which almost everyone in the community becomes concerned about long-term recovery.

We have concluded that recovery doesn't come about on a predefined timetable or schedule; it doesn't happen automatically; and it doesn't happen everywhere. Some communities, no matter how hard they try, never regain what they had before the event. Other communities, seemingly without much effort and sometimes in spite of themselves, become revitalized in short order.

Why, we wondered, do some communities do well while others languish or decline following an extreme event? Some economists have said that disasters actually stimulate the local economy because transfer payments from government and insurance companies, as well as the economic activity generated by rebuilding, largely offset losses from the event.[1] We weren't convinced. We saw communities where a significant share of the insurance payments and federal aid went to out-of-town and out-of-state contractors, and where the tumultuous effects on individual local firms were masked when one looked only at aggregate numbers in the metropolitan area. We saw whole communities divide their history into "before" and "after" the flood, hurricane, earthquake, or other extreme event. We saw businesses fold as a direct result of the event even five or six years afterward, when the owner had finally exhausted his or her savings, credit, energy, and hope. We saw one city on the verge of insolvency more than a decade after a hurricane had demolished it, despite having received tens of millions of dollars in state and federal aid and investments intended to enable recovery. We talked with individuals and couples whose lives were changed forever following a natural disaster, with one or both of them still getting psychiatric help and taking medication for long-term, clinical depression almost a decade later. They had lost loved ones or businesses built from the ground up to fulfill lifelong passions; their dreams had been shattered.

Some disaster researchers have suggested that recovery is pretty straightforward: one repairs, restores, rebuilds, or replaces the built environment and, before long, the community essentially returns to where it was before.[2] Again, we weren't convinced. We came to believe that recovery, if it means restoring things to what they were before the event, almost never happens. We came to appreciate that recovery comes about unevenly among sectors within the community. Local government sometimes recovers more quickly than the economy and vice versa. Some individuals recover quickly while others suffer intense anguish and depression for years. We have spent much of the last decade trying to understand why recovery is more difficult in some places than in others.

Over the decade, it became clear to us that the recovery challenge in any community was related to the nature and extent of the consequences of the extreme event on the community. By consequences, we mean much more than the direct and obvious damage to the community that can be seen by helicopter in the immediate aftermath. Systemic consequences often ripple through the community, unraveling some or all of its fabric. Because they dramatically complicate efforts aimed at recovery, these consequences are, in our experience, the biggest obstacles to recovery. Therefore, identifying and addressing the consequences that can destroy communities is the primary focus of this book.

INTRODUCTION

Not much compares with the intensity and anguish of having your possessions, your home, and your community collapse around you and those you love. Any disaster—whether flood, tornado, earthquake, or explosion—is devastating and remains with those who live through it for a lifetime.

For any one of us, a community-wide disaster is a rare event. Every year, however, across the nation and the world, scores of communities experience an extreme event—a shock sufficiently large to make the community and its local government unable to function as they did before the event, at least temporarily, and sufficiently ugly to have long-lasting adverse effects on almost everyone who remains in the community.

In the hours, days, and weeks that follow an extreme event, those who remain, and those who come to help them, have other powerful and unforgettable experiences as they work beyond mental and physical exhaustion to protect people and property, rescue survivors, recover bodies, and restore public services.

During that time, the injured are treated and the dead are counted; the water recedes and the mud dries, or the aftershocks become less frequent; blue tarps replace roofs that are literally gone with the wind; crews clean up foul debris and haul it away, volunteers scour farm fields to remove yesterday's treasures that would otherwise injure livestock and damage farm equipment; FEMA trailer parks appear; construction crews arrive; and rebuilding begins. At some point in that generally amorphous sequence, local officials and other civic leaders begin to think about and act in ways that they hope will result in the eventual recovery of their community. Local officials, already exhausted from days or weeks of living in city hall, face a set of activities and problems that most of them may never have seen before. Before they can turn their attention to their personal losses, they must address what they should be doing to restore economic, social, and political "normality" to the community—and to do so with a local government that is itself very likely a disaster victim.

We call this period after the cleanup and before the community regains its footing "the aftermath." This is when most communities experience adverse social and economic consequences stemming from the initial event. The term *recovery* is often used synonymously with rebuilding and repairing structures, restoring utilities, and cleaning up the visible effects of the flood, earthquake, or tornado. However, our research tells us that community recovery means much more and requires more than rebuilding physical artifacts. It is not "putting things back the way they were before." It does not follow as day follows night. Recovery, at the very least, means establishing viability within the post-event environment, viability for individuals and households, businesses, local government, and the community as a whole. It means adapting to new realities.

Some people think that recovery is simply a matter of cleaning up the debris and repairing or rebuilding the damaged infrastructure and buildings, but it is much more than that. The aftermath often brings unprecedented challenges to local officials and civic leaders. Apart from the injuries and deaths to residents and the damage to the built environment, these initial losses almost always lead to more complicated and damaging consequences for the community, such as major demographic changes, an unraveling of the local economy, and disruptions to social and political processes. It is easier to repair or replace the built environment, provided the community has sufficient resources, than it is to bring about recovery when the very fabric of the community has been torn.

Local government officials are rarely trained or prepared for the array of problems and challenges that unfold in the aftermath of disaster, and they often look to state and federal agency representatives for help. Unfortunately, while state and federal agencies and private organizations are skilled at providing disaster relief and helping to restore services and the built environment, they are neither responsible for nor proficient at long-term community recovery. Some federal agencies, such as the Economic Development Administration and the Department of Housing and Urban Development, provide financial support for projects intended to facilitate community economic development, and FEMA has funds to support initial response, cleanup, and rebuilding. But these agencies are not in a position to help with long-term community restoration. That part of recovery, arguably the most difficult part, is left largely to local officials who may be able to get help from state government.

This is when local officials, not knowing who else to call, reach for the phone to talk with officials in other communities about what they did and how they coped in the aftermath of an extreme event. Unfortunately, what worked in one community may or may not work elsewhere, so while the phone calls usually result in welcome moral support, the advice is not always as useful as might be hoped. Every disaster has its unique features and consequences because every community is a little bit different from every other one. And while it is possible to generalize across disasters to some extent, recovery is highly contextual: critical lessons must be extracted from individual disasters and viewed very carefully within the context of the character and needs of the specific affected community.

ACROSS THE NATION

In 1994, when we began our research on local government and disasters, we thought we were undertaking a relatively simple project in which we would identify and evaluate what local governments do after extreme events to help themselves recover from what are often staggering consequences. We planned to identify strategies that worked well and those that did not so that we could recommend to local officials how best to invest their time and resources to restore community viability and vitality. We soon learned, however, that our initial quest was, in fact, too simply defined. There is no "instant pudding" approach to "fixing" communities after an extreme event; again, what works is highly dependent on community context. Accordingly, we began to seek a deeper understanding of community recovery processes.

We studied each of our communities over a period of years, paying special attention to recovery processes in each, learning what each did to facilitate recovery, and assessing how those efforts panned out. Then we compared the communities' characteristics and their disasters. That approach helped us better understand long-term consequences and outcomes, identify events and patterns that were common to the recovery process or unique to a particular community, and learn what we could generalize to help those in other communities. In this book, we bring together what we learned from both victims of disasters and active participants in long-term recovery.

Work of this nature is inherently opportunistic. One goes where the disasters occur, not to some predefined, scientifically selected sample. Nonetheless, over the course of a decade, we were able to study a wide array of disasters in a variety of U.S. communities. We studied the consequences of earthquakes, floods, hurricanes, storm surges, tornadoes, and wildfires. We studied those consequences in communities in Alabama, California, Florida, Georgia, Louisiana, Minnesota, Mississippi, New Mexico, North Carolina, North Dakota, and Wisconsin. We spent time in more than two dozen communities, including larger metropolitan areas (Los Angeles and New Orleans), smaller metropolitan areas (Mobile, Alabama; Biloxi and Gulfport, Mississippi; and Pensacola, Punta Gorda, and Charlotte County, Florida), smaller cities (East Grand Forks and St. Peter, Minnesota; Grand Forks, North Dakota; Homestead and Stuart, Florida; Los Alamos, New Mexico; Montezuma, Georgia; and Tarboro and Rocky Mount, North Carolina), and small communities (Barneveld, Oakfield, and St. Nazianz, Wisconsin; Breckenridge and Florida City, Florida; Princeville, North Carolina; and Wahpeton, North Dakota).

In each community, we interviewed business owners and managers, civic leaders, residents, and political and administrative officials. To collect data we primarily used semistructured, face-to-face interviews, although we occasionally followed up with telephone calls. During each interview, we tried to begin without any assumptions about what had happened there immediately after the event and through the years that followed. We listened and tried to understand each story we heard from the respondent's perspective, and to purposefully interpret the various story elements in the context of our research purpose.

In our initial visits to each community, we typically started out in damaged areas by going into a business, nonprofit, or government office and asking the people we met, "Were you here when the disaster happened?" If they were, we asked, "What happened?" Then we listened to their stories, took notes, asked them to go into a little more detail on one or another particular point, and listened some more. Usually, they talked for an hour or so; some were done talking in 30 minutes, and others talked for several hours. Their stories provided us with insight into the decisions they made and the actions they took in their efforts to reestablish their businesses, communities, and lives; through them we were able to gain some understanding of each person's reality as he or she faced the challenges of recovery.

We have gone back to most of the same communities several times—for some, as many as seven times, which allowed us to review and validate or discount data we had collected during earlier visits. We interviewed many of the same people repeatedly at various intervals over three to seven years, learning how things changed for them since the event and coming to know many of them well. But with each visit, we added other people as well, mostly those employed by or elected to the local government at the time of the disaster or soon thereafter.

The stories people told us were accounts of activities and events conditioned by time. Over time, the stories we heard changed to some extent—in content, in perspective, or in context. At times, the latest stories conflicted with versions we had heard during earlier visits or with what others in the community had told us. These changes did not invalidate any of our data. Rather, they added another dimension what we had already collected. The intent of our interpretive process was not to settle upon scientific explanations, but rather to explain decisions, actions, and activities by relating their significance to the outcomes that ensued. Understanding the changes and related nuances as part of the data analysis challenged earlier interpretations and triggered alternative explanations.

In each community we listened to the assessments made by local officials, business owners, and community members to confirm what happened to the municipal tax base and finances, the local economy, and the population, and what, if anything, they did about it. Along with financial data, we also collected books, news articles, and other public records as available. These documents enabled us to triangulate information from the stories we had heard so that we could verify or add to them. Occasionally, information from one source conflicted with information from another, so we were required to determine which was more accurate by going back to our sources to seek reconciliation.

For our analysis we used virtually all the data collected from the interviews and the various documents we obtained. Over the decade we have identified and explored the complexity, nuances, and uncertainties that characterized the various communities' recovery processes. Our intent has been to capture and share practices that enhance municipal and community recovery in the short- and long-term after a disaster, and to offer interpretations that allow for further study and learning.

EXTREME EVENTS OF INTEREST AND CONCERN

Extreme events can result from a natural hazard event, an intentional or mindless act of destruction, a large accident, or a widespread virulent epidemic. They are, by definition, out of the ordinary, lying on the far end of the probability distribution for storms, earth movement, explosions, terrorism, and the like, and well beyond the range of what we expect to see in any given place in any given year. While they are rare in any given location, at least as measured by an individual lifetime, extreme events do occur somewhere on earth almost every day, more often in some locations than in others.

Natural Hazards

Extreme natural hazard events include moderate to great earthquakes, volcanic eruptions, and landslides; extended periods of extreme variation from expected weather patterns; moderate to large hurricanes, river and coastal flooding, storm surges and tsunamis; severe lightning storms, tornadoes, and extreme winds; and wildfires and urban conflagrations. Sometimes, extreme events are the result of interacting forces: tsunami as a consequence of earthquake, conflagration as a result of strong winds and drought, and so forth. The extraordinary tsunami of December 2004 was a consequence of a large earthquake off the coast of Indonesia.

Willful and Mindless Acts of Destruction

Willful and mindless acts of destruction have been the products of, among other things, uncontrolled violent mobs in Los Angeles, Detroit, and New Orleans; terrorist bombings in Oklahoma City and New York; the attacks on the World Trade Center in New York and on the Pentagon on September 11, 2001; fires deliberately set in dry brush in California; massacres at schools; and serial killers. Perhaps even more than natural hazard events, such acts of violence and destruction capture our attention as they diminish our sense of safety and well-being.

Large Accidents

Although they have been rare in the United States in the past half century or so, major industrial accidents are a persistent concern. The massive explosion of a ship loaded with fertilizer and small arms ammunition in Texas City, Texas, in 1947 was heard 150 miles away and killed hundreds of workers and onlookers. The toxic waste deposits at Love Canal posed massive health and cleanup problems. The partial meltdown of a nuclear reactor core at Three Mile Island, although it caused no deaths, had major social consequences— most notably, a long-term hiatus in the construction of nuclear power plants in the United States. The tanker truck crash that caused the meltdown of a major freeway overpass in Oakland, California, in May 2007, disrupted the commutes of tens of thousands of Bay Area residents for twenty-five days. The disruption would have been much longer had it not been for round-the-clock efforts to restore the freeway link. Railroad accidents involving the transport of toxic and hazardous materials are a continuing threat, and a large airliner carrying hundreds of people could crash in a crowded city on any given day.

Epidemics

We don't expect the bubonic plague to sweep across Europe any time soon, nor do we expect smallpox and other contagious diseases to sweep across the Americas, killing millions of people as they did following Columbus's arrival in the New World. However, there is serious concern about both the immediate and the long-term effects of the prolonged AIDS epidemic across Africa, of a pandemic influenza like the one that spread after World War I, and of other potential epidemics from new viral strains that seem to emerge every few years. Epidemics have shaped much of human history and are of constant concern as potential, perhaps likely, extreme events.

Inadvertent Side Effects of Conscious Choices by Others

There are also relatively slow-moving economic and social disasters. For example, Johnstown, Pennsylvania, where a great flood in 1889 killed more than 2,000 people, suffered another disaster when its steel mills shut down in the 1980s and 1990s. Elsewhere across the Rust Belt, individual communities, once prosperous and stable, watched helplessly as firms closed and moved elsewhere. Smaller towns built around Army posts and Air Force and Navy bases live and die depending on military base closing decisions. It is difficult to imagine, for example, how Lawton, Oklahoma, could survive if Fort Sill, home of the U.S. Army Field Artillery, or Watertown, New York, adjacent to Fort Drum and home of the 10th Mountain Light Infantry Division, were to close. Communities and local governments suffer devastating consequences when a principal or dominant employer closes its doors, or when an industry relocates elsewhere. Even though people looking in from the outside may not classify the loss of a community's economic raison d'être as a disaster, the consequences within the community are often comparable to those of a major flood, earthquake, or tornado, except that the onset may be in slow motion. For that reason, much of what we have to say in this book may be relevant to managers of communities that have experienced, or may experience, an economic disaster.

PLAN OF THE BOOK

This is neither a theoretical treatise nor a rehash of how to manage during the hurricane or flood. It is not a sermon on why everyone in the community should have paid more attention to hazard mitigation before the event or about how a city should prepare for the next. It is, instead, about what has happened to communities, what they did afterward, how it turned out for them, and what we learned from their experiences.

The first part of the book focuses on how extreme events become community disasters, how the consequences of extreme event can cascade through the community and beyond, and why some communities experience worse consequences and have a harder time recovering than others.

Part 2 focuses on the experiences of two dozen communities following extreme events. The events vary from one another, but they are all natural disasters. Likewise, the communities vary from one another, but all have in common their experiences with disaster and their attempts—sometimes, but not always successful—to recover.

Part 3 identifies the major consequences that we saw in cities across the country, whatever the cause of the disaster, and describes the response of local governments. We argue that a local government must recover sufficiently itself before it can be an effective agent of community recovery.

The fourth part of the book represents a compilation of what seems to help with recovery and what does not. We note the uniqueness of each community and the different paths to recovery that follow disasters, along with considerations and concerns that are common to almost all communities that experience a disaster.

PART 1

COMMUNITIES, DISASTERS, AND RECOVERY: SETTING THE STAGE

EXTREME EVENTS ARE NOT, IN and of themselves, disasters. They become disasters when they are perceived to have serious adverse consequences. The consequences that typically cascade through communities on the heels of an extreme event define both the disaster and the recovery challenge.

We have concluded that it is impossible to understand recovery without first understanding how adverse consequences ripple through a community following an extreme event. That is, it is important to understand the affliction before attempting to effect a cure.

This section is a summary of how we have come to see the consequences of extreme events on communities, what we have learned about what constitutes recovery, and what the pre-event characteristics of communities are that either facilitate or hinder recovery.

CHAPTER 2

WHAT MAKES FOR A COMMUNITY DISASTER?

"I would not give a fig for the simplicity this side of
complexity, but I would give my life for the simplicity
on the other side of complexity."

Oliver Wendell Holmes

EXTREME EVENTS ARE NOT, IN and of themselves, disasters. Extreme events become disasters when they have severe adverse consequences for people and the things that people care about. A hurricane's storm surge coursing over an unpopulated island might not be perceived as a disaster; however, if the same hurricane were to strike Miami Beach or Honolulu, it might well be described as a disaster, depending on the number of injuries and deaths, the extent of damage to the built environment, and resultant social and economic consequences for the local community. Maintaining a distinction between the event and the outcome facilitates understanding and analysis.

Clearly, adverse consequences vary according to the kind of event, its magnitude, the damages left in its wake, and the extent of those damages. Some outcomes might be called an "inconvenience" or an "emergency," whereas other outcomes might be called a "disaster," a "calamity," a "catastrophe," or even a "cataclysmic event." We contend that extreme events of the same magnitude—for example, a Category 3 hurricane—often result in dramatically different consequences for communities. Consequences vary in intensiveness, in extensiveness, and, importantly, in kind. They affect the probability that the damaged system will recover to something approximating what existed before the event or something better than what was. In terms of the human experience, each event creates a watershed for some people; however, some events affect many more people and are more significant than others.

DISASTERS: MAGNITUDE, CONSEQUENCES, OR BOTH?

The first reports following an extreme event usually include some scientific measure of its magnitude, whether it is the seismic measure of an earthquake, the wind speed of a hurricane, or the width of the swath cut by a tornado. Soon, however, public spokespersons and broadcasters begin to talk about consequences. They provide estimates of the number of lives already lost and the number of people who suffered injuries. This is usually accompanied by estimates of losses to the built environment: how many structures have been damaged and at what monetary cost, how many square miles are on fire, how far the river is above flood stage, where the levee failed, how many people are without power, and how many people have been displaced. These phenomena are easily observed, quantified, and understood.

However, understanding the full extent of losses from an extreme event requires going beyond an accurate body count, the number of structures damaged or destroyed, and estimates of how many cubic yards of debris had to be hauled away. It requires looking at less visible losses that are difficult to measure and evaluate, but that are incredibly important to community recovery. These less visible outcomes are also consequences of the event, but their impact may not be perceived until some time afterwards.

With few exceptions, built structures have social value only to the extent that they support the political, economic, and social activities and the historical heritage we define as important to our well-being. As structures yield to extreme forces imposed by earthquakes, high winds, or rushing waters, and as people are killed or flee, the stage is set for subsequent adverse consequences. These consequences ripple through the community and beyond, contributing to economic and social problems for those whose structures or communities may not even have been damaged by the event. Complex relationships within the community and between the community and other places are interrupted, damaged, and even ruptured by extreme events. Significant adverse consequences may threaten the continued viability of the community. The greater the amount of damage and the nature of the damage to social, political, and economic relationships, the less likely it is that the community will return to something approximating what it was before the extreme event.

COMMUNITIES AS COMPLEX, ADAPTIVE, OPEN SYSTEMS

The word *community* is used by many and can refer to any of several concepts. According to Morris Freilich, "since a requisite of science is specificity of terminology, we must conclude ... that at this time 'community' is a non-scientific term unless separately defined in every paper which uses it."[3] Robin Hamman employed George Hillery's analysis of ninety-four definitions of community, upon which we have based our use of the term.[4] According to Hillery, "The sociological term community should be understood here as meaning (1) a group of people (2) who share social interaction (3) and some common ties between themselves and the other members of the group (4) and who share an area for at least some of the time."[5]

We are generally comfortable with this definition, mostly because it is compatible with our own sense of a community as an open system, comprising individuals and institutions with patterned relationships among themselves and with the external environment. Although we acknowledge the use of the word *community* to refer to groups, such as a "community of disaster scholars," that may exist without the members living or working in physical proximity to one another but nevertheless communicating through a variety of means, our focus is on communities within a generally defined geographic space. Yet we also acknowledge that not every place that is called a community is actually a community, even by our generous definition.

We see cities, villages, small towns, and rural communities as complex systems comprising social, economic, and political patterns and relationships in a space usually occupied, at least in part, by a built environment. The built environment facilitates but does not constitute the functioning of the dynamic and interrelated system. Systems are often defined as a set of elements interrelated in such a way that the whole is more than the sum of the parts. Each person or organization is an element within the system. A disturbance to any one or more elements will have implications for other elements that make up the system.

Open systems interact with their external environment. In simple terms, they import information and "things" from the outside world, and export information and "things" to it. In a community, at least some people and organizations will have extensive relationships with the outside world. Local firms buy and sell in other communities, resources needed in the community come from other places, and the community provides needed goods and services to other communities.

The interrelationships between and among elements are generally patterned and persistent, producing a stability that helps define the community. At the same time, they are continually changing—although usually slowly and at the margins—as community members and institutions adapt to changes from both within the community and outside it. Communities maintain their stability through such continuing and collective adaptations, while with each new adaptation, the nature of the community as a whole changes. Comparing and contrasting snapshots of a community from year to year helps to demonstrate such change, not just in terms of buildings, streets, and parks, but in also in terms of who lives there, what they do, how they do it, and where they do it. Because the changes are largely evolutionary rather than revolutionary, they may not seem obvious to those who are fully immersed in the community, much as parents may not see major changes in their children that are apparent to outsiders.

We have come to believe that communities are largely self-organizing systems—that is, systems that adapt to change and increase in complexity through time without being guided or managed by an outside source. Individual and organizational behaviors can and usually do change over time. Those changes can take any of several directions and create any of several new trajectories for the community. A community system may survive an extreme event, but it will change to adapt to new realities. It might, for example, become smaller than the pre-event community or have significantly different demographic characteristics. Or the community system that emerges in place of the old one may actually be "bigger" or "better." However, change is not always for the better. The new community may be inferior to the old one in any number of ways. The economy may not be as strong, a historical district might be destroyed, or the post-disaster population may have fewer marketable skills that the pre-event population.

A disaster is a big jolt to a community system, one that punctuates the equilibrium that the community has achieved. A community that suffers such a jolt often finds itself in serious trouble, facing an uncertain future. If the extreme event is sufficiently large, individuals, organizations, and institutions may find it difficult, even impossible, to perform critical functions and maintain important relationships in the aftermath. This has significant, but not easily anticipated, effects on relationships among the various elements in the community, as well as on their relationships with the outside world. New relationships emerge as individual elements of the system change behavior in an attempt to stabilize themselves in the new, post-event environment. These elements, in their struggle to establish or reestablish relationships with other elements within or outside of the damaged system, usually begin by trying to do what they did before the event with the assumption that the disaster was "just a bump in the road." Some adapt to the new realties and survive, and some do not. Being able to cope successfully with the consequences of an extreme event is not a given.

If one looks at the community from this perspective, it becomes easier to understand how the system as a whole can suffer consequences from an extreme event even when parts of it do not sustain direct physical damage from it. Those consequences are examined in this chapter.

UNFOLDING DISASTERS

Disasters unfold episodically in fits and spurts as the extreme event, acting like a catalyst, triggers consequences, and those consequences, in turn, trigger other consequences. Some extreme events, such as hurricanes, may provide warnings. Others, such as earthquakes, do not. When they strike, extreme events have significantly different durations: some last but a few seconds (e.g., earthquakes) while others may drag on for weeks or months (e.g., flooding). The initial consequences are injuries and death to some inhabitants of the stricken area, damage to some features of the natural environment, and damage to the built environment. Other things being equal, we think that the greater the damage to the natural and built environments and the more deaths and injuries proportionate to the pre-disaster population, the more likely it is that cascading consequences will result from the event and recovery will become more difficult.

As we studied the diverse community experiences following extreme events, two points were hammered home. The first is the importance of the consequences that ripple out from the initial event on recovery processes and outcomes. The second is that it is difficult for anyone to anticipate the full array of outcomes from the event. Some might say that what happens in the aftermath is a simple matter of cause and effect and, as such, should be predictable, but we are not convinced of that. Such thinking presumes a linear chronology of discrete events in which one cause leads to one or more effects, and so on.

Our observations suggest something different: that multiple causes generate multiple effects along varying timelines. Consequently, the post-disaster chain of events is less of a chain and more of a cascade of seemingly diffused events, many of which may be interrelated. Consequences interact with one another, unanticipated relationships appear, individuals make choices that may or may not be surprising, and some phenomena may simply proceed randomly. We believe that it is virtually impossible to reliably predict before an event all the major consequences of that event. One can anticipate some of the consequences and prepare accordingly, but no matter how well prepared a community may be, some consequences will arise that no one will have anticipated. At that point, it may be more important for the community to focus on its ability to adapt rather than on its ability to predict the future.

Some observers have suggested that extreme events simply accelerate existing trends in a community.[6] For example, perhaps 20,000 older middle-class people who had depended on the defense industry left the Northridge, California, area after the 1994 earthquake. Some were already planning to leave before the earthquake struck; the event may have simply hastened the exodus. In other examples, such as Montezuma, Georgia, and Grand Forks, North Dakota, the central business district was declining because retail development at the edge of the city or in larger cities nearby was drawing a greater share of business. After extreme events in 1994 and 1997, respectively, left these cities flooded, the decline in downtown businesses apparently accelerated. We agree that extreme events often accelerate preexisting trends, but we are inclined to believe that they can also produce discontinuities and altered trends. While it is generally possible to sketch out the broad picture of what is likely to happen in or to a community following an extreme event, it can be difficult to predict the outcomes of most extreme events reliably. Too much "prediction" is hindsight bias in action. Saying something like "I knew that would happen" doesn't suggest that one is a good predictor so much as that one is able to piece together events after the fact.

Even as consequences continue to unfold, attempts at recovery are made at many levels in the community. Individuals and households struggle to regain some approximation of what they perceive as normal; so, too, do businesses, nonprofit organizations, and governments. Their collective efforts constitute their attempts at community recovery. Eventually, a "new normal" develops for each element in the community. The community system, if it survives, changes and moves on.

CHAPTER 3

THE CASCADING CONSEQUENCES OF EXTREME EVENTS

"You can observe a lot by watching."

Yogi Berra

The remains of Hurricane Katrina in southern Louisiana. Photograph taken in March 2008.

ENGINEERS USE TERMS LIKE *CASCADING* failures, *progressive collapse,* and *sequential collapse* to describe phenomena associated with the failure of certain kinds of infrastructure, such as buildings, bridges, or complex machines. In some cases, failure of one structural element leads to the failure of other elements. Sometimes the linked, sequential failures end only when the structure fails completely, such as when a bridge collapses into a river. Other times, the sequence of failures is attenuated for some reason, such as when the structure includes "fuses" that limit sequential failure; in those cases, some of the structure remains intact or at least minimally functional after the dust settles. Essentially the same thing happens in communities, except that communities are much more complex systems than are bridges, buildings, or machines. Moreover, while bridges and buildings may be "dynamic" in terms of motion and response within limited ranges, they do not change through time to alter the nature of those dynamics and the relationships among their parts. Communities do. Nonetheless, the notion of cascading or progressive damage is useful in illustrating what happens in communities after an extreme event.

The sequential consequences of an extreme event are affected greatly by phenomena both within and outside the community. Sometimes those consequences seem to radiate out almost seamlessly, like the ripples in a pond when a stone is dropped into it. Other times, it is difficult to distinguish them from the immediate effects of the event itself. Was the collapse of the twin towers of the World Trade Center the immediate result of the impact and explosion when the airplanes flew into them, or was it an outcome of a long chain of events, starting with the towers' design and construction? For most of us, it doesn't matter because the sequence of the most salient events occurred within a short period, and we think of it as a single, continuous event.

Still other times, the consequences are not as readily apparent as ripples in the pond. The fact that thousands of small businesses went out of business in

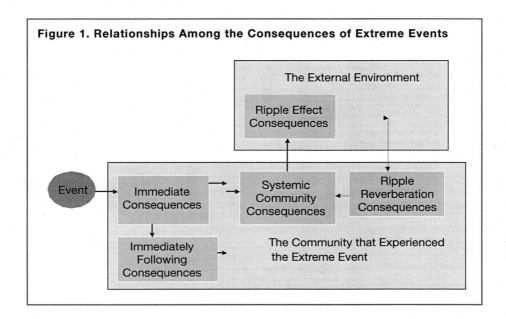

Figure 1. Relationships Among the Consequences of Extreme Events

the immediate area of the World Trade Center because their customer bases collapsed along with the towers was also a consequence of the attacks, although not as visible or visually dramatic. Similarly, some workers and volunteers who descended on the site to help in the recovery and cleanup effort were exposed to the dust and smoke left by the explosions, fire, and collapse and now appear to be suffering with chronic illnesses from that exposure.[7] We therefore think it useful to distinguish among different kinds of post-event effects in order to facilitate the development of more effective mitigation and recovery strategies. To this end, we have devised a straightforward classification scheme, shown here in Figure 1 and Table 1, to clarify the differences in consequences.

In the schema shown in Figure 1, an extreme event triggers both *immediate* and *immediately following consequences* for the community: buildings crumble or are flooded; people are injured or killed. Then, the very direct effects result in consequent events: building contents become water-soaked because an earthquake triggered the water sprinklers; fire ravages the city because the water pipes have been broken and firefighters are unable to fight the flames. After that, the initial consequences lead, almost inexorably, to *systemic community consequences:* with jobs gone and housing unavailable, at least temporarily, people relocate—either temporarily or permanently; pre-event relationships are damaged or destroyed, so the community cannot function as it did before. Depending on the extent of those disruptions, a critical mass of relationships may remain or be reestablished to help put the system back together in short order so that it can approximate the pre-event community, but that is not guaranteed. Recovery may require establishing a working set of relationships in the aftermath of the disaster.

We also identify *ripple effects* and *ripple reverberations*. Following Katrina, Baton Rouge became Louisiana's biggest city overnight, as thousands of individuals and dozens of businesses rushed to make it their home, at least temporarily. After the earthquake that destroyed Kobe, Japan's main harbor, other harbors on the Pacific shore of Asia were able to capture much of the shipping that had gone to Kobe. A ripple reverberation consequence is that Kobe has not been able to regain its pre-event preeminence as an Asian seaport.

These consequences and examples of them are elaborated upon below.

Table 1. A Classification of Consequences Following from Extreme Events.

Type of Consequence	Definition
Immediate consequences	The initial effects of an extreme event on life and the built or natural environment.
Immediately following consequences	Effects that are tightly coupled and causally related to the immediate consequences, usually having an impact on the built or natural environment.
Systemic community consequences	Social, political, and economic outcomes that emerge in a geographically, socially, politically, or economically defined area that are triggered, exacerbated, or ameliorated by immediate and immediately following consequences of an extreme event.
Ripple effects	Consequences that "ripple" out from the area that suffered the event and that affect other places.
Ripple reverberations	Consequences for the community that suffered the initial disaster because ripple effects lead to changes in other communities and those changes subsequently ripple back to the place that experienced the initial event.

IMMEDIATE CONSEQUENCES

Immediate consequences are the first results of the collision between an extreme event and the built or natural environment. Immediate consequences include injuries and deaths, as well as damage to physical infrastructure, buildings and their contents, vehicles, and the natural environment. Immediate consequences are highly visible: as a direct consequence of the January 1994 earthquake, an apartment building in Northridge collapsed, killing a score of people; whole villages were swept away by the 2004 tsunami in the Indian Ocean;[8] bridges and buildings—some buildings up to six blocks away from the shore in Gulfport and Biloxi—were smashed by Katrina's storm surge in 2005; homes in California are consumed annually by wildfires; and almost every year, a heavy snow load from an unusually severe storm collapses a roof on a commercial or industrial building somewhere in the United States. These are all immediate consequences. So, too, are people killed or injured by the force of a blast, the rush of floodwaters, the strength of tornado winds, a rioting mob, or any other extreme event. People have been killed or injured when falling from buildings or through windows during an earthquake; others have lost control of their vehicles during the shaking and have crashed and died. Again, these are all immediate consequences.

IMMEDIATELY FOLLOWING CONSEQUENCES

Immediately following consequences are triggered by immediate consequences. For all practical purposes, immediately following consequences exist within a generally traceable cause-and-effect relationship.

Examples of immediately following consequences abound. Perhaps the most frequently cited example in the disaster field is the great fire that followed the 1906 San Francisco earthquake. The earthquake damaged buildings, and that damage spawned one or more fires. The earthquake also damaged the water supply system so badly that firefighters could not use it to quell the blaze; soon, virtually the entire city burned. In New Orleans, immediately following consequences of Hurricane Katrina include the flooding that occurred in part because the levees could not stand up to the hurricane's storm surge, and the deaths that resulted from inadequate life support (e.g., not enough food, water, or health care for people who needed it, when they needed it) in the aftermath. Other examples include

- People dying because a hospital has been destroyed by an earthquake and they cannot get treated for a preexisting illness or an injury unrelated to the earthquake
- People dying of heat exhaustion because a widespread power failure during an extreme heat wave has left them without electricity to run fans or air conditioners
- Damage to building contents (e.g., carpeting, furniture, files) because the shaking from an earthquake triggered water sprinkler heads to go off or ruptured the rigid water couplings to toilets, causing water to flow for hours and soak everything inside.

The incidence of immediately following consequences for people appears to be a function of the extent to which people depend on others for help. The greater the reliance on others who may also be affected negatively by the disaster and the fewer the options for outside help, the greater the likelihood that immediately following consequences will occur. High dependence on relatively few tightly linked resources increases people's vulnerability.

The incidence of immediately following consequences for physical structures appears to be a function of how various elements of structures and infrastructure are connected to one another—how they are coupled with the physical environment within which the damage occurs—as with, for example, the rigid connection between the water supply and a toilet's water tank. Most buildings in the United States now have fuses and circuit breakers to keep electrical overloads from creating a subsequent fire. When there is no "fusing," slack, or shutoff valve between vulnerable community or building components that are connected to one other, the likelihood of immediately following consequences is greater.

SYSTEMIC COMMUNITY CONSEQUENCES

Systemic community consequences follow both the immediate and the immediately following consequences of an extreme event. As they consist of the post-event unraveling of some or all of the community's social and economic infrastructure, systemic community consequences alter the community context when the event occurs, dramatically complicate efforts aimed at recovery, and thus, in our experience, may be the biggest obstacles to community recovery.

Systemic community consequences follow from the actions that people and organizations take after the initial emergency period in response to some set of stimuli, including but not limited to immediate and immediately following consequences. Systemic community consequences are numerous, pervasive, and extremely important—for example, business closings and consequent unemployment after a flood or earthquake. Businesses usually close at least temporarily (or fail to reopen immediately) after an extreme event because they have to clean up and check for damage. Repairs may be necessary. But if a firm's real property assets are destroyed, the business might be closed for an extended period of time, perhaps even forever.

However, even businesses that suffer little or no damage often find themselves in trouble after an extreme event. When losses from immediate and immediately following consequences are significant, residents change their spending priorities. Some will have lost their primary source of income and, if they have put any money aside, may begin to draw down their savings. Spending patterns change with changes in priorities. Repairing one's home and replacing lost contents become high priorities. Firms that sell building materials and household appliances may do very well after a disaster, but firms that sell goods and services, the acquisition of which is optional or deferrable, may suffer. Such businesses include optometrists, upscale restaurants, dry cleaners, most kinds of specialty shops, and others that rely on the discretionary income of local residents and visitors. Such firms lose customers for extended periods, their owners and employees lose income, and their suppliers lose contracts. Many of these establishments struggle to remain afloat by drawing down on owner equity and by obtaining loans they will not be able to repay unless their customers return.

Systemic community consequences are often difficult to analyze because they stem from multiple events or conditions that may be hard to identify and assess in terms of importance to the community. WLOX-TV (serving Biloxi, Gulfport, and Pascagoula) reported, following Katrina, that Oreck, the vacuum cleaner manufacturer, was closing its plant in Harrison County (Gulfport area) and moving its operations to Tennessee.[9] Oreck reportedly cited "the increased cost of doing business and the harsh realities of living on the gulf coast" as its reason for moving. Local officials, however, suggested that Oreck's departure might have been triggered more by the fact that the tax breaks it received for coming to Harrison County ten years earlier would expire at the end of 2006.[10] The latter impetus may well have contributed to the former.

Following extreme events, employment opportunities typically emerge in temporary or new businesses, including cleanup and construction. Local firms may benefit from increased sales of building materials and supplies. However, when people are insured against the hazard or have sufficient credit, most of the homes in a damaged area get new plumbing, new carpeting, and new drapes within months; after that, it is a long time before they will need to make such purchases again. This sometimes creates significant challenges for smaller merchants in the neighborhood that specialize in building supplies and home furnishings.

Housing prices almost always change in a community after an extreme event—another systemic community consequence. Prices change when the event has greatly depleted the housing stock and housing in the area surrounding the disaster site is insufficient to meet the demand. Housing prices did not soar in Northridge, primarily because most of the Los Angeles metropolitan expanse suffered little or no damage and could absorb the displaced residents who wanted to stay in the area. On the other hand, local officials in several coastal communities in Florida reported that housing costs increased rapidly following recent hurricanes because housing in the immediate area became so scarce. Housing shortages are followed by rent increases, often putting the housing out of reach for lower-paid workers.[11]

People are often surprised to find high housing prices after a disaster. At first blush, people assume that housing prices will be lower because "who wants to live in a disaster zone?" In general, however, property that was desirable before the disaster will continue to be desirable unless the root causes of the desirability are altered, and property that is desirable is expensive. Thus, land values in some coastal areas have tended to increase dramatically in the wake of some hurricanes; over the past decade or so, Hurricanes Ivan, Dennis, Katrina, Rita, Wilma "cleared" many land parcels along the Gulf Coast shoreline. Large price increases have followed an extreme event even though the hurricane or other storm literally blew housing off concrete slabs just months before. The homes and businesses swept away tended to be older and smaller, and the vacant land became available for bigger, more modern homes and condominiums. Small motels were replaced by large hotels; small businesses were replaced by larger firms. Some areas almost look as though they underwent instant urban renewal.[12]

People often move away from the area after an extreme event, some expecting to return and others not. People who were not residents sometimes move in, often making significant changes to the community's demographic characteristics. In Northridge, California, and the southern part of Dade County, Florida, the new residents had less education, lower incomes, and fewer marketable skills than those who left forever. In Port Charlotte, Punta Gorda, and other coastal cities along the Gulf of Mexico, new residents tended to be older and wealthier than those who left. As the demographics change, so too do other aspects of the community. For example, as the voters change, so do their preferences and their priorities; political agendas are rewritten, and different issues move toward the front burner. The changes may not be immediately obvious, but every community changes to some extent after a significant extreme event.

It must be noted, however, that two communities that experience almost identical immediate and immediately following consequences after an extreme event may experience dramatically different levels and kinds of systemic community consequences; this is because they had significantly different social and economic characteristics before the event. Some communities with extensive immediate and immediately following consequences sustain relatively few systemic community consequences. We puzzled over why several small Wisconsin villages we studied recovered quickly from tornadoes that devastated between 25 and 40 percent of their homes and businesses, but resulted in only a few deaths and injuries. The communities were rebuilt to be very similar to what they were before the tornadoes hit, except that the rebuilt areas had younger, smaller trees and newer, more contemporary-looking housing.

We concluded that the communities recovered quickly because they suffered so few systemic community consequences. Each of the villages developed originally to serve area farms in the horse-and-buggy era, but autos, trucks, modern highways, phone service, modern plumbing, and efficient winter snow plowing made them largely redundant as farm support communities. In the decades following World War II, they transformed into bedroom communities ten to twenty miles from an urban center. The communities consisted mostly of members of a few extended families. In each case, there were few deaths from the tornado because siren warning systems were adequate and because community members became first responders in advance of and in support of public safety officers. Almost all the homeowners and businesses were well insured against tornadoes, and even though the damage was extensive within the community, it did not stretch the area's construction capacity, so residents were able to rebuild quickly. Moreover, since many residents were retired and others worked in the urban center that was about a half hour away by car, few residents lost their jobs. Thus, the community itself did not suffer extensive damage; only the buildings were destroyed.

RIPPLE EFFECTS

We define ripple effects as those consequences of an extreme event that manifest themselves outside the community. Except in unusual cases, ripple effects have limited impact on the disaster-stricken community; instead, they affect the rest of us. It is important to acknowledge these consequences because they are often overlooked in state and national response to disasters and in recovery support efforts.

Hurricane Katrina provides ample illustrations of ripple effects. Almost immediately after the hurricane and subsequent flood, Baton Rouge became

Louisiana's largest city. In the two weeks after Katrina, 400,000 square feet of office and retail space were leased in the city.[13] Public school enrollment increased 9 percent, and Catholic school enrollment increased 25 percent. In the following weeks, traffic volume increased 37 percent, traffic accidents increased 19 percent, and arrests increased 25 percent. The initial population increase in the Baton Rouge area was an astounding 57 percent, but six months later it had settled down to somewhat less than a 25 percent increase from pre-Katrina levels. While recovery and support efforts were focused on New Orleans and the surrounding parishes, Baton Rouge was left to deal with the ripple effects almost entirely by itself. The city's hospital capacity had been strained, but as of February 2006, Baton Rouge hospitals had yet to be reimbursed for medical services provided to New Orleans evacuees.

Local governments elsewhere experienced ripple effects as well. Even as far away as Wisconsin, welfare and health care costs increased because hundreds of New Orleans evacuees were relocated there. Midwestern grain farmers had grave concerns about being able to ship their harvest because much that was intended for foreign markets was scheduled to go through the Port of New Orleans. Fortunately, the port did not suffer severe consequences, and grain shipments were handled fairly effectively.

Ripple effects are not necessarily negative, however. Following Katrina, Las Vegas experienced a significant increase in conferences, conventions, and tourists that otherwise might have gone to New Orleans.

RIPPLE REVERBERATIONS

The fifth type of consequences from extreme events—ripple reverberations—occur in disaster-stricken communities when ripple effects subsequently come back to haunt them. Kobe, Japan, fell in importance as a Far Eastern port after the 1995 earthquake because other ports that had been competing with the city took advantage of Kobe's misfortune to increase their own share of the Far East shipping market. Kobe may never regain its preeminence as a Far Eastern ocean shipping port, having fallen from third in the world to eighteenth in tonnage in the decade after the earthquake.

Following the 1900 hurricane in Galveston, Texas, nearby Houston—further inland and perceived to be considerably safer—began to grow rapidly. In 1902, just two years after the devastating hurricane, President Theodore Roosevelt approved federal funds to build a ship channel from the Gulf of Mexico to Houston. By the time the channel was completed in 1914, Houston had surpassed Galveston in size and economic importance. Galveston's developmental trajectory had been changed permanently. This is not to imply that the different developmental trajectories of Houston and Galveston were entirely a consequence of the hurricane's effects on Galveston; for example, we cannot know the extent to which the federal government's grant for the shipping channel was a consequence of people's reluctance to rebuild in Galveston or of the discovery of oil at Spindletop in Beaumont, near Houston, in 1901. The oil trade greatly enhanced the need for a Gulf port at the same time that Houston became preferred by investors after the hurricane killed more than 6,000 people in Galveston.

We believe not only that ripple effects and ripple reverberations may be as important to a community's post-event well-being as systemic community consequences but also that they sometimes exacerbate systemic community consequences.

IMPLICATION FOR RECOVERY

The immediate and primary implication of this approach to understanding the range and diversity of consequences from extreme events is straightforward: simply repairing, rebuilding, or replacing damaged structures and facilities is necessary, but not sufficient, to effect recovery of a community that has experienced systemic consequences. Despite the quote from the movies, even if you build it, they might not come.... back, that is.

CHAPTER 4

RECOVERY: BEYOND RESTORING THE BUILT ENVIRONMENT

"The world is full of people whose notion of a satisfactory future is, in fact, a return to the idealised past."

Robertson Davies, "A Voice from the Attic"

A former Days Inn along the Mississippi Gulf Coast after Hurricane Katrina swept most of the coastline clear. Photograph taken in March 2008.

W E ASKED MANY PEOPLE FROM varying backgrounds and from all across the country to tell us what they meant by *community recovery*. We listened as federal, state, and local officials; local business owners and managers; local residents; practicing engineers and professors of engineering; social scientists; and emergency management personnel responded. Their answers revealed a lack of consensus, although almost everyone living in a community that had experienced an extreme event used the term recovery as though we all share a common understanding of what it means.

Some people described recovery in terms of physical manifestations: recovery occurs when the debris is cleaned up, when little or no physical evidence of the disaster remains, or "when the FEMA trailer park is gone." Sometimes people talked about recovery in terms of administrative manifestations: "Recovery is done when the recovery projects are completed and the emergency offices are closed. FEMA and the SBA [Small Business Administration] are gone." Others talked in terms of restoration or replacement of structures, facilities, and services: "Recovery wasn't complete until I got electricity back and that was more than four months after the hurricane." For still others, recovery was defined in terms of pre-event and post-event levels of service. We heard people talk about a community as having recovered when its infrastructure was rebuilt, and its municipal and public services were back to pre-event levels.

Still others relied on simple community statistics. Is the municipal population back to pre-event levels? Is the tax base back to where it was before the event? Some economists say that recovery has occurred when aggregate economic activity is back to what it probably would have been in the area had the event not occurred in the first place. Individual business owners we spoke with often said their recovery would be complete when business activity and profits had returned to pre-event levels. Housing officials seemed inclined to define recovery as that time when people got out of temporary housing and into permanent housing.

People use the word *recovery* to mean many different things, and that creates situations in which analysts often "talk past one another." Since there are neither negative consequences for this lack of shared understanding nor apparent positive consequences for reaching a shared understanding, the imprecision in use and meaning continues. *Recovery* generally has two meanings relevant to this discussion: (1) to bring back to a normal position or condition (e.g., to recover one's balance after stumbling) and (2) to regain a normal position or condition (e.g., to become healthy again after an illness). In the case of a community that has experienced an extreme event, it is too easy to interpret these definitions as suggesting that recovery occurs when the community is restored to *status quo ante calamitus*—that is, the same way things were before the disaster. We suspect that a community can never be restored to exactly what it was before the disaster. After all, communities change and evolve through time, regardless of whether a disaster strikes. On a micro level, people move in and out of the community, babies are born, and people die every day. On a macro level, membership in organizations changes along with changing demographics. Some businesses are starting up as others are ceasing operations. New people win government posts while others retire or move on. Accordingly, no community is exactly as it was a year previously. Thus, it doesn't make practical sense to talk about recovery as restoring everything to what it was. That's returning to a past that is already over and done with; it is not recovery.

A NEW PERSPECTIVE

If community recovery were simply a matter of restoring utilities and repairing or rebuilding damaged structures, it would be a relatively easy matter to have things back in good order within a relatively short period of time, provided that enough money were pumped into the community. If that had been the case, then Homestead, Florida, would not have been nearly as insolvent as it is a decade after Hurricane Andrew; Montezuma, Georgia, would have a thriving central business district; Princeville, North Carolina, would be a model community; and the Mississippi Coast and communities in Louisiana hard-hit by Katrina would be "recovered" as soon as someone found the wherewithal to rebuild the buildings that had been swept away or flooded and consequently demolished. But communities are more than their buildings and utilities. Buildings and utilities are a means to an end, not an end in themselves.

We believe that communities, especially those considered "effective" by a majority of observers, are largely self-organizing systems that consist of many interacting parts. They develop their structures largely as a function of interactions between and among the system's components (e.g., schools, employers, local government, churches). While affected by influences external to them, each self-organizing system is unique in that its structure depends on the nature, frequency, and reaction to the interactions of the internal system. Self-organizing systems are inherently democratic: no all-powerful figure (e.g., government) determines all actions, reactions, and consequences. No one part of a self-organizing system controls the whole system, or even any part of it, without being affected or constrained by other parts of the system.

If we were to peek into a self-organizing system to which we did not belong, the system might well appear chaotic and incomprehensible; this is because we wouldn't be able to observe all the interactions taking place or even understand many that we *were* able to observe. Peeking into an unknown self-organizing system would be analogous to Alice trying to figure out Wonderland. Self-organizing systems usually make sense to their inhabitants, although not necessarily in a way that inhabitants can articulate to an outsider.

The sum of behaviors by individuals and organizations following a disaster can take several directions and create any of several new trajectories for the community. The pre-event community system may survive, but with changes that are responsive adaptations to new realities. The community might, for example, be smaller or have significantly different demographic characteristics. Or the community system that emerges in the place of the old one may actually be "bigger and better." Change is not always for the better, of course. The new community may be inferior to the old one in any number of ways: socially, economically, politically, and so on.

One of the considerable challenges to post-disaster community recovery is that people generally dislike ambiguity, and they particularly dislike it when it follows on the heels of a disaster. After one loses a loved one, one's home, one's job, and/or one's sense of security, there is already enough ambivalence. People seek some security in that turmoil, and they often see it in what used to be (at least as they recall it), so they want to return there. Of course, they can't return, but it usually takes months for them to understand that. Sometimes it takes years.

WHAT IS COMMUNITY RECOVERY?

The challenge of community recovery is defined by the nature and extent of the problems generated by the collision between the pre-event community and the extreme event itself. Community recovery begins when the community begins to act as a self-organizing system in which myriad interactions yield a shared sense of community. We have come to understand that recovery is relative; there is no fixed point at which recovery can be said to have taken place. Community recovery, to us, is occurring as a community becomes capable of developing through time as a generally self-sufficient entity within the generally accepted social, economic, and political standards of its regional and national context. Recovery has happened when the community repairs or develops social, political, and economic processes, institutions, and relationships that enable it to function in the new context within which it finds itself.

When communities are viewed as open, self-organizing systems, as comprising not only structures but also people and organizations and the relationships among them, and when the implications of cascading consequences in the community system are understood, it becomes clear that recovery requires much more than simply restoring the built environment.

The establishment of viability in the present and for the future is the critical variable that defines community recovery. Viability in the near future means that the community has a developmental trajectory projected to result in continued self-sufficiency and that its key institutions are coping with and adapting to changing circumstances. We believe, too, that there is no recovery unless the community is generally acceptable to a critical mass of the residents and is congruent with generally accepted standards within the region and the nation.

Recovery emerges as the outcome of several sets of activities: restoring basic services to acceptable levels, replacing infrastructure capacity that was damaged or destroyed, rebuilding or replacing critical social or economic elements of the community system that were damaged or lost, and establishing or reestablishing relationships and linkages among critical elements of the community. The recovered community may closely resemble the pre-event community in many ways, but it need not. The extent of recovery should not be measured by how closely the post-event community resembles the pre-event community.

Recovery Is Never Assured

Until the New Orleans post-Katrina debacle, even veteran students of disaster recovery sometimes talked as though post-disaster community recovery were a certainty. We suspect that this optimism reflected a strong desire to maintain control in the face of ambiguity, and a continuing belief that with the right tools and attitude, anything that has been broken can and should be fixed. This thinking is consistent with the belief that human beings can control nature or, at a minimum, are capable of anticipating and dealing with whatever nature dishes out.

In fact, history suggests that post-disaster community recovery is not a certainty. One need only reflect briefly on the city of Pompeii and its total inability to recover from the catastrophic eruption of Mount Vesuvius to know that recovery is not guaranteed.

Some local government officials see their job in recovery as limited to getting the infrastructure and local services back to some proportion of pre-event capacity. That view appears to be premised on the belief that the rest of community recovery will fall into place relatively automatically thereafter or that it is someone else's responsibility to see that it does. Communities that experience adverse consequences to the local economy often begin trying to attract new businesses while providing little or no help to existing businesses in trouble. Why? Because local officials often have lots of experience trying to attract new businesses and almost none trying to help those that already exist. This is not to say that some local governments do not think and step "outside the box" after a disaster in an attempt to bring about community recovery. They do. But many do not, sometimes because they don't believe that they can.

There Is No "Recovery Timetable"

Many people we talked with a year after a disaster struck their community seemed surprised that the recovery was not complete. When giving a presentation in spring 2008 on Hurricane Katrina and its impact on New Orleans, one of us was asked whether New Orleans was recovered, and if not, when it would be. In trying to answer the question, the presenter noted that "New Orleans" comprised several distinct neighborhoods, economies, and sociocultural elements, some of which resembled their pre-Katrina states and some of which, for better or for worse, did not. Local officials who promise recovery in six months or a year are doing their communities a serious disservice, setting up expectations that simply cannot be met. Recovery cares little for the calendar.

Some communities recover from extreme events much more quickly than others. Some of the places that recover quickly are the beneficiaries of prudent choices and investments made by community leaders both before and after the event. Others seem simply to be in the right place at the right time. Long-term changes in markets and regional competitive advantage affect the rate and nature of post-event recovery. No matter how skilled local leadership may be or how much money is available, recovery is much more difficult when a community is in an area suffering from a long-term decline in competitive advantage. Another influential factor is the extent to which the pre-disaster community was one holistic entity rather than a collection of contiguous entities—like New Orleans.

This suggests that, at least in part, community recovery is a complex phenomenon with multivariate antecedents or causes. It also hints at the roles played by pre-existing conditions in the community and by communities outside the focal community. A community does not operate in a vacuum, and its past will condition (but not dictate) its present and future.

Furthermore, and as suggested earlier, while a community may "recover," there will be individuals, families, and organizations that do not recover—even many years later—in the sense of returning to their former lives. We interviewed people for whom, years after a flood, earthquake, or tornado, the nightmare had yet to end. Some had lost a child to the event; others had lost their marriages. Still others had lost their homes, their jobs, and their sources of income. Neither individuals, nor neighborhoods, nor organizations recover at the same pace or in the same pattern. To expect otherwise is to ignore the fact that people and organizations have different "starting positions," make different choices, and face different microenvironments—the specific environments that affect them individually or organizationally.

CHAPTER 5

FORCES THAT COMPLICATE COMMUNITY RECOVERY

"In these matters, the only certainty is that nothing is certain."

Pliny the Elder (23 AD–79)

Small businesses inevitably suffer. Photograph taken in New Orleans in October 2005.

HE CHALLENGES ASSOCIATED WITH LONG-TERM recovery vary considerably among communities. One major variable is the nature and extent of the damage to the built environment and the deaths inflicted directly by the extreme event. A second variable concerns the extent of dysfunctional systemic effects on the community itself and the impact of any ripple reverberation consequences that might result. In this chapter, we explore both of those variables, as well as of some others. But first we consider an important question: recovery for whom and to what extent?

RECOVERY: AT WHICH LEVEL AND FOR WHOM?

When disaster strikes, it can, and usually does, occur at many levels. Similarly, recovery from that disaster can take place at some levels and not necessarily at others.

The consequences of a disaster always affect individuals and families or households. As individuals, we may suffer physical or emotional injuries. Someone close to us may be injured or killed. We may lose our jobs or homes, or we may experience some other great financial loss. Our dreams may be shattered, and we may be unable to rebuild or replace them. Each of these consequences can be tragic.

Similarly, neighborhoods, communities, organizations, municipalities, and sectors of the economy may experience significant losses. In a sufficiently large event, such as the rapid succession of Hurricanes Katrina and Rita, regions may suffer, and as the effects ripple out still further, national and global economies may suffer significantly as well.

Recovery can take place at the individual level without resulting in community recovery, and vice versa. We have spoken with people who left a devastated community to seek a new life in a different place because they saw that as the surest way for them to achieve personal recovery. Similarly, recovery at the community level does not ensure that the individuals in the community will have recovered. Illustratively, one might observe that Community X has rebuilt its infrastructure, has a thriving economy, a growing population, and a bright future, but it may be that many or most of the people living there before the event have gone elsewhere to work on their personal recovery and that those who comprise the population now are new arrivals.

In this book we focus primarily on recovery of the "place"—the area within the boundaries of the local government where the extreme event occurred. In so doing, we consciously use the term *community* and use it somewhat loosely. We know, of course, that a large city might comprise several communities and that a single community might be divided by municipal boundaries. Nonetheless, we believe it is the most useful term available to us for this book.

THE NATURE AND EXTENT OF DAMAGE TO THE BUILT ENVIRONMENT

Two fundamental dimensions of damage to the built environment affect the recovery challenge: (1) what got damaged or destroyed, and (2) how badly it was damaged or how much of it was destroyed. A simple biological analogy is useful here. If a large rock falls on your head, your recovery challenge will be more difficult than if it had been a small rock falling on your foot.

Some parts of the community are not easily and quickly replaced. Losing one of those parts is usually a severe blow to the overall community. For example, virtually the entire central business district in Montezuma, Georgia, was flooded and remained so for several days. Many residents lost jobs they held in the central business district, and shop owners there lost income for an extended period. However, only a few houses in Montezuma were damaged, and the people who lived in them found other places to shop, mostly out of town. Most of them did not come back to the downtown businesses when the central business district was repaired, beautified, and reopened.

We think it is probably easier for a community to recover if the damage is widespread but not particularly devastating to any particular geographic area or to any sector of the community, such as employment, housing, or retail. That is, if given a choice between light, extensive damage and heavy damage that affects and is contained within one entire sector of the community, we would opt for the lighter, more extensive damage. Our experience suggests that it would be easier to recover from a flood that puts a little water in everyone's basement than one that destroys the central business district and leaves housing intact.

If only a fraction of the housing in a community is damaged or destroyed, residents can usually find different or temporary housing, but when a community loses most of its housing, the problem becomes considerably more complex. In the Northridge earthquake, comparatively few homes or manufacturing facilities were destroyed. As has been noted, there was adequate housing for everyone, even though it may have been miles from where residents had lived previously. And while the earthquake devastated the Northridge Fashion Center, other shopping malls in nearby areas were not damaged and thus benefited from the extra business. Northridge, as a place, recovered fairly quickly, even though the demographics of the community changed significantly. By contrast, with 80 percent of New Orleans under water and inaccessible for weeks after Hurricane Katrina, people had nowhere to live, shop, work, go to school, and so on. Nearly three years later, much of the city's housing is still damaged to the point of demolition, schools remain closed, and businesses remain shuttered.

PRE-EVENT COMMUNITY CHARACTERISTICS

We have become convinced that pre-event characteristics of a community have a significant bearing on whether community recovery will occur. If the pre-event community has anticipated and taken steps to reduce the direct impact of an extreme event, it will suffer less direct damage when the event occurs, other things being equal, and, presumably, less in the way of systemic damage. And if the economy is strong, if the community is a magnet that attracts people from other areas to make it their home, and if the economic base is primarily in growth industries, economic and demographic recovery is markedly more likely.

But prospects for recovery from an extreme event are greatly reduced when there are extensive losses to the built environment in a community that is already vulnerable because of conditions that existed before the event. Again, a biological analogy may be useful: a bout with influenza is an inconvenience to a young, otherwise healthy person, but it may kill a person who already suffers pulmonary problems or is weakened by other diseases. People understand and relate to this instinctively with regard to humans, but relatively few bring that understanding to bear on communities. In essence, the starting position for a community facing recovery *after* an extreme event depends, to a considerable extent, on the community's standing *before* the event.

Exposure and Vulnerability

Losses to the built environment are primarily a consequence of the magnitude and duration of the event coupled with the exposure and vulnerability of critical elements within the community. The city of Pensacola, Florida, received less damage than did Pensacola Beach from Hurricanes Ivan in 2004 and Dennis in 2005. Pensacola Beach is a barrier island unprotected from winds and storm surges, whereas the city of Pensacola is protected by both the barrier island and a coastal hill that runs roughly parallel with the shore. In the case of earthquakes, given the same ground motion, California will suffer far less damage than a city in Indonesia, the Middle East, or even the central United States, simply because the buildings in California are built to standards that anticipate earthquake damage. Accordingly, they are less vulnerable.

For construction in flood channels and floodplains, Wisconsin's Department of Natural Resources enforces state regulations that exceed federal standards. As a result, communities along Wisconsin's rivers typically suffer much less damage during flooding, other things being equal, than similar areas in neighboring states, and much less damage than they would suffer with less stringent regulations.

The lesson is clear. If you reduce exposure of housing, utilities, critical facilities, and employment centers when it makes sense to do so, you reduce the vulnerability of that which remains exposed. Not doing so almost guarantees more adverse consequences and a longer and less probable recovery.

The Local Economy

Communities with a strong and diversified economic base typically recover more quickly and easily than communities with a faltering economic base, other things being equal. A community's economic base might be strong for any of several reasons. One reason is location: communities located near a special and in-demand resource base not found in many other places tend to have a strong economic base. The resource might be an exceptional labor force, an abundance of minerals not found many other places, outstanding recreational or aesthetic features, or miles of sandy beaches in a place that is warm and sunny most of the year.

The kind of employers in the community also makes a big difference in recovery. Communities are fortunate when they are home to firms that are in growth industries, that are well-positioned within those industries, and that are committed to the community for economic or other reasons. When a firm loses its productive facilities, owners must choose between reinvesting in the community or investing somewhere else. If they decide to invest elsewhere and not repair or replace their damaged facilities, the community's recovery becomes more difficult. But if they are willing and able to make major investments in the community in the aftermath of an extreme event, the likelihood and pace of recovery increase substantially.

We are inclined to believe that firms owned by longtime community residents are more likely to rebuild and reemploy locals than are businesses, factories, and employment centers owned by national or international conglomerates to whose management the community is an abstraction. In the latter case, those interested in having the facility rebuilt will have to demonstrate to the owners that corporate investment in that place is more likely to produce the desired financial return than investment in some other location.

Communities with economic problems before the extreme event typically experience significant systemic community consequences. Some of these communities, like Homestead, Florida, lost the community's major employer, Homestead Air Force Base, because of a decision made in Washington, outside the community, just before Hurricane Andrew. Homestead community leaders told us that they also believed the community suffered from the North American Free Trade Agreement (NAFTA). The area had been a major truck farming area before NAFTA, supplying fresh vegetables and fruit to the Miami region and across the nation, but NAFTA, they said, moved most of the truck farming south to Central America, which compounded the recovery problems after Hurricane Andrew hit.

San Francisco was almost completely destroyed in 1906 by earthquake and fire, but it rebounded quickly—likely because it has an extraordinary harbor located in a temperate climate at the mouth of rivers draining large, prosperous agricultural areas. Santa Cruz, badly damaged in the 1989 Loma Prieta earthquake, also rebounded quickly because it is in a great location with a pleasant climate and has a large and steady employer, the University of California–Santa Cruz.

Similarly, communities along the Gulf Coast have done reasonably well after a spate of hurricanes in the years just before Katrina and Wilma. Biloxi appears to be recovering well from Katrina, thanks in part to the economic engine of its casinos, which employ more people now than they did before Katrina.[14] Pensacola recovered to a considerable extent because, as has been noted, so much of it is behind a big hill and thus was not damaged; in addition, its tourist trade is relatively stable except during hurricanes, and it has many amenities, including a warm climate and great beaches. But Escambia County has not fared as well because the bulk of the county is located north of Pensacola, away from the beaches in somewhat less desirable space, and is dependent on agriculture.

COMPOUNDING EFFECTS

Our studies strongly suggest to us what is probably obvious: communities that suffer minimal damage are likely to recover much more quickly than communities that suffer large losses to an extreme event. And strong communities are more likely to suffer fewer adverse systemic consequences than are weak communities, other things being equal.

The matrix shown in Figure 2 depicts our hypothesized relationship between two variables—the preexisting conditions in a community and the extent of damage to the community—and the probability that the community will recover. The figure depicts a hypothesized relationship between the consequences of an extreme event on various kinds of communities and the likelihood of recovery. Again, we have concluded from our case studies that recovery and the time required to achieve recovery is never a given. We believe that rebuilding and restoring the physical artifacts of a community might follow a generalizable timeline, but recovery of the community system and of its local government does not.

Figure 2. Hypothesized Probability of Community Recovery from an Extreme Event.

		Community Status before the Disaster		
		Vibrant, strong	**Viable**	**Highly vulnerable, weak**
Degree of Disaster Consequences	**Modest**[1]	Highest probability of recovery		
	Moderate		Likely to recover	
	Severe			Lowest probability of recovery

[1]Consequences may be immediate, immediately following, ripple effects, or ripple reverberations

RESOURCES

The chances of a timely recovery are greatly diminished when there are inadequate resources to help with rebuilding public and private infrastructure and reestablishing the necessary economic and social relationships. Extensive resources are needed to rebuild homes and facilities and to help bring about long-term social and economic recovery. In the United States, there will be federal and, most likely, state aid as well as charitable giving from across the country. Depending on the nature of the event and the residents' wherewithal, there will be insurance payments to both residents and local governments. Without that influx of insurance payments to homeowners and firms, recovery will be slow and painful.

Disaster-struck communities that expect money to be delivered in packets of cash for them to use however they deem appropriate are doomed to disappointment. So, too, are communities that expect the relief and recovery funds to arrive within a few hours of a massive disaster. The money always comes in more slowly and with more strings than anticipated.

Money is not the only critical resource. Effective leadership is essential. Elected officials who put personal agendas before community needs, display excessive partisanship or factiousness, are corrupt, or are given to rhetoric rife with hyperbole do nothing to facilitate long-term recovery and may well delay it. Along with competent leaders, community recovery requires competent, dedicated public employees.

MULTIPLE DISASTERS AT ABOUT THE SAME TIME

For any community, the succession of multiple disasters in short order usually exacerbates the spread of systemic community consequences. It was bad enough for Homestead to have the Air Force base there close, but then to have a 35-mile-wide swath nearly flattened by Hurricane Andrew was almost the final blow.

A community suffers a double blow when an extreme event strikes in the midst of a sagging economy. Montezuma, Georgia, had been having economic difficulties for some years before the central business district sustained serious damage from a major flood and could not fully reopen for more than a year. Similarly, Orleans Parish had serious social, political, and economic problems long before it was flooded in the wake of Hurricanes Katrina and Rita. Whenever cities are in trouble before an extreme event, our experience indicates that they will have a very difficult and painful recovery, if they recover at all.

FORCES AFFECTING SELF-ORGANIZING BEHAVIORS

Just because communities are self-organizing doesn't guarantee that the community system that emerges in the aftermath of an extreme event will be desirable. The emergent community will be desirable only to the extent that a critical mass of its members are willing and able to take actions that contribute to system reorganization rather than to continued dissolution. At the extreme, if everyone chooses to leave and no one replaces them, the community will no longer exist.

Consequently, one of the forces that can complicate recovery is the dominance of dysfunctional or nonfunctional characteristics in a community. We have no idea how to alter those characteristics in any politically palatable way; all we can do is point out that local social and political culture will affect the nature and pace of recovery.

Social and Political Culture

Part of how communities respond to disasters has to do with the dominant political culture of the area and the state within which the community exists. When studying Grand Forks, North Dakota, and East Grand Forks, Minnesota, we found distinctly different attitudes toward recovery even though the communities are barely 200 meters from one another. The State of North Dakota provided relatively little assistance to communities in the Grand Forks area that were flooded; most of the assistance to rebuild came from the federal government. By contrast, the state of Minnesota was extremely active helping communities in the flooded areas along the Red River of the North, even providing technical assistance and money to East Grand Forks so it could replace its lost housing through a program that had the local government build a subdivision and create a home-building program. Some people in North Dakota thought the Minnesota approach "smacked of socialism." To those in Minnesota, however, the approach seemed like a pragmatic way to deal with tough problems.

Political culture influences choices about whether to have building codes, how rigorously to enforce them, and to what extent the local community should innovate, imitate, or lag behind. When building codes are not enforced rigorously, when infrastructure is not well maintained, and when communities don't innovate, chances are very good that the built environment will be vulnerable to extreme events. What local officials can do is constrained by what is deemed appropriate within their political context. We have seen, too, that the level, extent, and duration of cooperation, mutual assistance, and charity varies among communities.

Collective Efficacy

"Collective efficacy" refers to a group's shared perception about its ability to face challenges and overcome obstacles. We are interested in the extent to which collective efficacy varies—not by group but rather by community. We have been unable to find research that addresses the question directly, but we have the sense from our work that some communities have higher collective efficacy than others. The apparent results of differences in collective efficacy include the strongly held and shared belief that the community can and will recover, and the subsequent speed with which people in the community regain their footing and tackle the problem of recovery. For example, we have found small northern communities that cleaned up from tornados before state officials and FEMA arrived. One consequence of their collective efficacy was that they lost out on federal financial assistance for cleanup. At the same time, the people in these communities felt prepared to deal with whatever might happen in the future, with or without assistance from outside parties (e.g., state and federal government agencies).

In general, we think that community recovery is much less onerous when a majority of the residents believe that they are preeminently responsible for determining what happens to the rest of their life. In other words, communities most likely to recover see themselves as self-organizing, not reliant upon some external agency acting as a patronizing "father figure." High collective efficacy means not waiting for bailouts or special help and not believing that others "owe" the community its recovery.

PART 2

POST-DISASTER EXPERIENCES: WHAT HAPPENED IN OTHER COMMUNITIES

*T*HIS SECTION AND THE FOLLOWING one focus on the experiences of two dozen communities after extreme events. The events vary, but they are all natural disasters: floods, hurricanes, tornadoes, earthquakes, wildfires. The communities we studied vary, too. Some are large and some are tiny. Some had strong economies; others did not. Some were highly effective communities before the event; others were not.

We have tried to pull together and aggregate, where appropriate, what we think are the most important lessons from each disaster and each community. We hope the reader learns as we did from the experiences of real people living and working in real communities.

CHAPTER 6

OFTEN THE LOCAL ECONOMY UNRAVELS

"It was a 360-degree disaster: home, job, business, every aspect of life."

Homestead, Florida businessman looking back at Hurricane Andrew.

"I think we'll be ok. I got a job cleaning restrooms."

Flood victim who lost his business.

A formerly thriving service station. Photograph taken in New Orleans in October 2005.

LOCAL OFFICIALS HAVE TO BE concerned with maintaining a vital local economy to ensure that municipal revenues are adequate to meet the needs and demands of residents.

The local economy starts to unravel when a sufficient proportion of firms in the community suffer losses to their production, sales, or distribution capacity—and particularly when they are unable to restore that capacity quickly. Unraveling also results from damage to a particular resource on which the community's economy largely depends—for example, attractive sandy beaches and warm tropical waters. Over the past few years, hurricanes in the Gulf of Mexico erased homes, hotels, and rental accommodations from barrier islands and oceanfront property, depositing the debris on the white, sandy beaches and into the water. When the beaches are not usable, tourists stop coming. Capacity is also limited when hotels, motels, and restaurants are damaged or demolished. Undamaged businesses may be affected as well if they cannot ship products; if power, water, and sewerage are not available; and if suppliers or customers are unable or unwilling to do business with them.

Workers disappear, at least for a while, as the storm approaches and after the earthquake strikes. For the first few days or weeks, they are busy taking care of their families and their immediate needs. However, if a substantial amount of housing is damaged or destroyed, many of them leave, at least until the white FEMA trailer parks are put in place days or weeks later. Workers also leave if they cannot find work that matches their skills or expectations. This sometimes ends up in an ugly spiral: workers are not there because there is no work, and employers do not reopen because there are not enough workers.

If a company's suppliers are damaged or cannot get the supplies to the company, and there is no convenient substitute for those supplies, the firm cannot produce its goods or services. Unless it has large inventories, the firm may have to close—at least temporarily. The same holds true when the firm loses its customers because they have either moved away or changed their spending priorities.

HOW THE LOCAL ECONOMY CAN UNRAVEL

In virtually every community we've studied, the local economy unraveled to a greater or lesser extent following the disaster in ways that were unique to each community. For our purposes, the local economy consists of two basic parts. One part comprises those businesses whose primary customers live in the neighborhood or in the community—mostly smaller businesses that provide services and support to the community, such as dry cleaners, groceries, restaurants, pharmacies, gas stations, real estate agents, and furniture stores.

The second part includes the businesses that drive the local economy and create significant linkages with the outside world. Often larger employers than those in the service industry, these businesses might be engaged in manufacturing, mining or other extractive industries, regional distribution, transportation, and financial and insurance services. They may also include private utilities, health care organizations, educational institutions, and elements of the tourism industry.

Smaller Businesses and Nonprofit Organizations Serving the Local Community

There is only a weak relationship between the amount of damage to its building or inventory that a business sustains following an extreme event and that business's subsequent recovery. Some small businesses that suffer considerable damage rebound quickly; others with minimal damage never bounce back. We have seen small businesses fail as many as half a dozen years after a disaster as a direct consequence of that disaster.

If a business cannot provide what its customers want when they want it, it is in serious jeopardy. Customer loyalty doesn't go very far when a need is urgent. The customer usually goes somewhere else, especially if it is not particularly inconvenient to do so. In Northridge, California, a local carpet dealer lost most of his inventory because the earthquake set off the sprinkler system in his warehouse, and then heavy rains poured more water into the building through the broken walls and roof. While he was trying to get the landlord to repair the building he leased, large manufacturers from Georgia and distributors from elsewhere in California cut deals with the big contractors, offering truckload prices and immediate delivery. Like other small retailers, he finally acquired suitable space and an inventory just in time to learn that there was no longer a need for his product or services Virtually all the immediate demands for carpeting had already been met, and when most of the buildings in the market area get new carpeting all at once, there is little demand for it for several years thereafter. With the replacement carpet business essentially gone for several years, this businessman had to rely on growth in the area, which proved insufficient to keep his business afloat. His business folded two years later.

If a business loses a critical mass of its customers for an extended period and cannot replace them, it is in trouble. Businesses lose customers for several reasons. First, some customers move away after the disaster and may not come back. Second, customers' priorities change: the new priorities have to do with repairing damage and rebuilding, perhaps replacing home furnishings and an automobile. At least for a while, discretionary money dries up, and people do not buy expensive new fishing gear or expensive jewelry, or eat in expensive restaurants, and they may postpone their annual or semiannual eye exams as well as nonessential work on their cars. Businesses that provide goods or services that are purchased with discretionary income usually make a mistake if they rush to reopen. Their insurance companies will urge them to reopen, citing unfounded statistics about how many businesses fail if they do not reopen immediately following a disaster. However, if the business has business interruption insurance, the insurance will stop paying as soon as the firm reopens, regardless of whether that business ever has another customer or makes another dime.

In nearly every community we studied, small-business owners struggled to reopen only to learn that they no longer had customers. A long-flourishing upscale fishing-tackle shop found that no one was interested in buying high-end fishing gear in the wake of the disaster. Hobby shops, jewelry stores, specialty restaurants, and other small firms experienced the same thing. Some closed shortly after they reopened. Others tied to a lease or an unrealistic dream that things would return to the pre-event normal, kept pouring their savings into a dying business until they were out of savings, credit, energy, and hope. Some years after a flood, we interviewed an elderly woman who had been in business for many years with her husband. They had nothing left but the building in which the dying business was located. Only half in jest, she asked, "Do you know how to find someone who will start a fire?"

Instead of reopening right away, it makes sense for the owner to assess whether there are still customers for the business. If not, it makes sense either to move to a new location where customers did not suffer as much loss or to shift to another business that is in demand. An optometrist in Northridge chose the first option: after cleverly sending out a "please return if not delivered" letter to all his patients and receiving 40 percent of the letters back, he decided to move to an undamaged area and rebuild his practice. Others chose the second option. We learned of a woman in a Gulf Coast city who was in the catering business; since there was no one to whom to cater following the hurricane, she mothballed the business and went into the temporary storage business. A couple with a cooking school saw no one come through the door for some time, so they opened a small restaurant to meet the growing needs of recovery workers. The last we heard, both entrepreneurs were doing just fine and meeting important needs in the community as it struggled toward recovery.

The extent to which an individual business can recover in the aftermath of a disaster seems to depend on three variables. First, it must be able to recover assets lost in the extreme event. This usually means having the right kind of insurance in the right amounts from a reliable insurer. Most of the small businesses we studied had inadequate coverage, no coverage, or the wrong kind of coverage. Only about 11 percent of those we studied following the Northridge earthquake had earthquake insurance. The proportion covered by flood insurance in other communities was not much higher. All too often we were told, "I didn't think I needed flood insurance. I'm in the 500-year floodplain." Apparently the insurance agents for these businesspeople failed to tell them that a large proportion of the annual losses to floods in the United States are *in* the 500-year floodplain. Among those who were covered by flood insurance, most had inadequate coverage. Many of those who were insured in areas struck by hurricanes became embroiled in conflict with their insurer over whether damage was caused by wind or water. Water damage was typically not covered by anything other than flood insurance.

Second, a small business's recovery depends on the extent of adverse effects on its suppliers, employees, and customers. Unless the small business has access to all three of these critical ingredients, recovery is impossible. Suppliers provide the raw materials (inputs) needed by the business. Employees transform those raw materials into finished goods and services (outputs). Customers purchase the finished goods. The money received from customers is reinvested in the business, enabling the business to purchase needed raw materials and hire employees. This cyclical process enables the business to interact effectively with its environment and survive.

Third and finally, a small-business owner must be able to adapt quickly and appropriately to new realities in the post-event environment. Suppliers with whom the business enjoyed long-standing relationships may have gone out of business or left the area, or may have stayed but now need to renegotiate pricing in order to remain viable. New suppliers may need to be found; and their expectations for contract negotiations and inventory management may be radically different from those of the former suppliers. Employees may have left the area (perhaps because their homes were destroyed, or their children's schools were destroyed, or their partner's employers went out of business) or may have stayed but now demand higher compensation because of increased opportunities. As was the case with new suppliers, new employees' expectations for compensation, working hours, job requirements, and so on may differ from those of former employees. Longtime customers may have moved away, and new customers may demand better, faster, cheaper, and easier products or services; their needs and expectations having little or nothing in common with those of former customers.

Trying to do the same old thing usually doesn't work. Often, the owners need help in understanding this. The ones that do understand have a better chance of recovering.

Businesses That Drive the Local Economy

Historically, many businesses in American communities were owned by people who came from the community, started the business, grew it, and passed it on to family and other trusted colleagues. These individuals had a real stake in the community. They were attached for reasons that went beyond short-term profit–and-loss statements. They contributed to the economic integrity of the community in myriad ways: through employment; through the attraction and then support of complementary industries; and through basic support of community members, both as individuals (e.g., with scholarships) and in the aggregate (e.g., with financial support of community events). The names associated with these businesses were found throughout the community: on buildings, parks, streets, and so on. For many reasons, this model of doing business in one's local community is no longer dominant.

The local company producing frozen ice cream treats that was owned by a local family is now owned by a huge international conglomerate. It didn't take long for the conglomerate to move 400 premium jobs halfway across the continent "to achieve greater efficiency." Most of the shareholders in the huge tissue manufacturing firm across the river once lived in the community, but the firm went public and now the shareholders are all over the world. The packing plant on this side of the river was owned by a local family, but was bought and operated by a large national firm and is now owned by a large foreign firm. If the production facility is damaged or destroyed, the decision of whether to repair or rebuild the plant will be made far away by people who may never have been to the plant, who do not know the employees, and who may not even know where the community is. The decision will be based on what is best not for the community but for the shareholders and managers of the conglomerate. If it makes financial sense to the firm's decision makers to continue in the damaged community, the business will be rebuilt; otherwise, it will not.

The decision about whether to rebuild operations in the community depends on many things: replacement costs, shipping costs, market shifts, and production costs. It depends, too, on the availability of an adequate labor force. If the workers have left the community and cannot return because there is not enough housing or because it is too expensive for them, that will factor negatively into the equation. However, if the firm has limited relocation options, has the prospect of being able to operate profitably, or is tied to the community by virtue of its mission, it may provide temporary housing for workers. Faced with a housing shortage in New Orleans after Katrina, several hospitals committed to the area chose to house their employees rather than lose them to other cities.

Communities that are home to a large and presumably permanent institution—a university, military base, a Veteran's Administration hospital, or the like—benefit substantially from the continual flow of money from the state or federal government to support that institution. The resulting jobs and income help sustain them after extreme events. And, unlike the case with a large for-profit firm, the local community may be able to exert pressure on its state legislature or Congress to stall or preclude any closure (e.g., of the military base) that might be considered.

CONDITIONS UNDER WHICH UNRAVELING IS MORE LIKELY

Although an economy will not necessarily unravel following a disaster, in most of the communities we studied, it did. The lesson seems clear: it is not enough to rebuild the infrastructure and the buildings. We have identified the following factors that, from our research, appear to be correlated with a greater likelihood of economic calamity after an extreme event.

- *Massive damage and massive systemic effects.* When there is massive damage to the community's physical artifacts, and where the social and political systems become dysfunctional, we believe that economic recovery will be difficult and time-consuming.

- *A weak pre-event economy.* When the pre-event economy in the community and the surrounding area is in trouble and has been for some time, we believe the economy is more likely to unravel. By "in trouble" we do not mean that unemployment is up because of a nationwide recession. Rather, we mean that the community's economy has serious, long-term structural weaknesses. For example, some communities that began as service centers for an agricultural area are no longer needed for that purpose; larger centers and good highways have made them superfluous. Cities in the Rust Belt that depended on manufacturing steel and heavy machinery can no longer do that as the industry moved away.

- *Nonlocal owners.* From the communities we have studied, it appears that the local economy has a better chance of rebounding when the larger employers are locally owned. Owners with local ties seem more likely to reinvest in the community than are owners who are unfamiliar with the community except as a line on a spreadsheet they review quarterly.

- *Inadequate quality or quantity of workers.* In one community we visited several times over the years, a local told us that the community had serious labor problems: "Chances are, if they can read, they can't pass the drug test." An untrained, numerically inadequate labor force will not support economic recovery. Firms that were in the community before the extreme event occurred will be seeking another location with an adequate workforce, even if wages are lower in the disaster-stricken community, other things being equal.

- *Inadequate infrastructure.* If the local infrastructure was marginal before the disaster and indications are that it will not improve, we would expect business owners and managers to seek an alternative location. It may be as close as the suburbs or a nearby community near a major highway, but it could be in another place altogether.

- *Increased costs of doing business.* Sometimes the consequences of a disaster increase the costs of doing business. If housing becomes scarce and prices rise, workers may become scarce and wages may rise. Delays in construction may increase the cost of doing business for only a few years, but that may be enough to induce the firm to seek a better location.

- *Heavy "penalties" for rebuilding.* Johnstown, Pennsylvania, has been badly damaged by floods on more than one occasion. During one particularly destructive recent flood, steel manufacturing facilities were severely damaged. One condition of rebuilding the facilities was that they would have to meet contemporary environmental safety standards. The steel firms balked at what they perceived to be extraordinarily high costs. They did not rebuild the steel mills, and Johnstown's economy declined precipitously.

- *A deteriorated location relative to other places.* Relative changes often have as much impact as absolute changes. One area can become less desirable simply because another neighborhood has been created that is far more desirable. Heavy public investment in interstate highways, recreational areas, universities, research parks, airports, or other important facilities can influence the relative desirability of a location. And a community that has benefited from public and private investment in infrastructure and amenities is likely to realize economic recovery faster and more completely than one that has not benefited from such investment.

CHAPTER 7

POST-EVENT DEMOGRAPHIC CHANGES

"I've developed a new philosophy... . I only dread one
day at a time."

Charles Schulz's Charlie Brown

One of many schools that did not re-open, largely in response to population
changes. Photograph taken in New Orleans in February 2008.

OFTEN, BUT NOT ALWAYS, EXTREME events lead to population
changes in both the stricken community and those communities to
which people have fled for safety. The changes show up in the total
number of people living in the community and their demographic characteristics. After a disaster, some people leave and other people arrive. Some leave
temporarily because their home is destroyed and their job is gone; many of
them, at least initially, plan to return. New people move in shortly after the
disaster to help clean up, restore services, and rebuild, and then they usually
leave. Others come to live and work in the community as it rebuilds. Following
an extreme event there are usually disjoints between the number of people

and the availability of housing, labor, and services. Communities are complex, so it's no surprise that matching demand and supply across a wide range of interests and activities does not proceed smoothly or evenly.

TOTAL POPULATION

Every disaster-stricken community has some immediate and temporary changes in the number and kind of residents. People whose homes are badly damaged or destroyed make arrangements to live somewhere else—down the street, across town, or out of town, depending on individual circumstances and options—until they can find more permanent housing. After Katrina, displaced New Orleans residents who were unable to return to their homes for weeks or months moved everywhere. Some moved to nearby Jefferson Parish to the west, or across the Mississippi River to the south. Some moved to Baton Rouge, 80 miles to the northwest. Some moved to Houston, Texas, 350 miles to the west. Others moved all over the United States.

Sometimes the move is intended to be temporary. When damage is widespread, everyday life becomes burdensome and sometimes unaffordable. "I took the kids and moved in with my sister in [insert name of town] until the power was back on and the supermarket reopened. My husband stayed there to [help clean up, to work, because he was a fireman]." "My apartment was destroyed and I couldn't find an affordable place to live so I moved to [insert name of town] where I got a job [waiting tables, at a hotel, driving cab]. I'll come back [when I can find a place I can afford to live, when I can get my old job back, in a few months]." A woman in Minnesota moved away after the tornado destroyed her home: "I moved to a house trailer in [a city about fifteen miles away], but I continue to work in the family business. It's been almost ten years and I still can't find housing in [her home town]. I don't know when I'll be able to move back."

Other times the move is intended to be more or less permanent. Thousands of middle-class homeowners left Northridge after the 1994 earthquake with no plans to return. Many were at or near retirement age. The area's defense industry had seen cutbacks, so some of these people had been laid off or offered early retirement packages. But taxes were high for retirees, and the value of homes had softened or actually declined. For many, the earthquake was the last straw. In Homestead, Florida, of the many people who left just before or just after the hurricane struck in 1992, a large number were Air Force personnel, their dependents, and retired career military people who lived near the base to avail themselves of medical services, base exchange, and commissary privileges. Others were retired "snowbirds" who left because almost everything was destroyed and other places beckoned.

Unless they have strong economic or emotional ties to their community, one warm and sunny Florida community might be perceived to be about as good as another. Moreover, some communities are less likely to be destroyed by a hurricane than others, at least in the minds of the residents or evacuees. We interviewed some younger people who left Florida simply to live and work elsewhere. "It wasn't all we had hoped for, and the hurricanes are simply too much to deal with. There are other nice places that don't have them." These individuals considered themselves to be mobile and acted accordingly.

In New Orleans, population estimates continue to be fraught with error, in part because many people continue to move in and out and around within the city, so the population is in flux. And because much of the city's housing stock remains damaged, homeowners who are unable to return to their damaged homes full time may still spend considerable time at their properties, repairing them or preparing them for repair or sale. Other homeowners continue to live with family and friends in the city or nearby parishes while they wait to fix, sell, or demolish their homes and start anew. As of March 2008 (two years and seven months after Katrina), 71.8 percent of the households in the city were actively receiving mail.[16] The rate of repopulation appears to be slowing, however, and few believe that New Orleans will return to its pre-Katrina population within the next decade or so.

It is useful to look at the experience of communities that experienced disasters to see how their respective populations changed from before and after the event. Table 2 presents total population numbers for sixteen communities. The data suggest several important conclusions. First, total population of a community almost always declines in the period immediately following the disaster. Second, although total population almost always declines immediately following a disaster, regional and sectoral trends tend to dominate over the longer term.

With some exceptions, populations in rural towns and cities in what the Census Bureau calls the North Central States are either stable or declining.

Table 2. Population Change in Communities Having Suffered Disasters from 1990 to 2006 and Percentage Change Between the Last Census Count Before the Extreme Event and 2006 (estimated).

Event Year	Name of Community	Type of Extreme Event	1990	2000	2006 (est)	Percentage change between last count before the extreme event and 2006
North Central						
1997	Grand Forks, ND	Flood	49,425	49,321	50,372	1.9
1997	Breckenridge, MN	Flood	3,708	3,559	3,357	(9.5)
1997	East Grand Forks, ND	Flood	8,658	7,501	7,857	(9.3)
1998	St. Peter, MN	Tornado	9,421	9,747	10,693	13.5
Atlantic Coast States						
1994	Montezuma, GA	Flood	4,506	3,999	3,973	(11.8)
1999	Princeville, NC	Flood	1,652	940	1,668	1.0
1999	Rocky Mount, NC	Flood	48,997	55,893	57,057	16.5
1999	Tarboro, NC	Flood	11,037	11,138	10,564	(4.3)
Florida and Florida Gulf Coast						
1992	Florida City, FL	Hurricane	5,806	7,843	9,445	62.7
1992	Homestead, FL	Hurricane	26,866	31,909	53,767	100.1
2004	Punta Gorda, FL	Hurricane	10,747	14,344	17,126	19.4
2004	Charlotte County, FL	Hurricane	110,975	141,627	157,536*	11.2
Gulf Coast: Louisiana, Mississippi						
2004	Pensacola, FL	Hurricane	58,165	56,255	53,248	(5.3)
2004	Biloxi, MS	Hurricane	46,319	50,644	44,342	(12.4)
2004	Gulfport, MS	Hurricane	40,775	71,127	64,316	(9.6)
2005	New Orleans, LA	Hurricane	496,938	484,674	223,388	(53.9)

*2005 estimate

The 1990 and 2006 total population counts for Grand Forks show about a 2 percent growth; however, from 1990 to 2000, the decade within which the city was flooded, the population declined slightly. St. Peter, which is in growing, economically sound eastern Minnesota between the Twin Cities and Mankato, grew 13.5 percent from 1990 to 2006, despite having almost 40 percent of its homes damaged or destroyed in 1998. Both East Grand Forks and Breckenridge, Minnesota, smaller cities along the Red River of the North, both show population declines of almost 10 percent in the decade and a half. The small towns we studied in the Southeast declined in population or remained essentially stable. Montezuma, Georgia, and Tarboro, North Carolina, lost population over the decade and a half embracing the flood years. Princeville lost nearly half its population after the 1999 flood, but, largely thanks to the construction of large public housing projects, had regained its 1990 population by 2006.

Cities in Florida, including those on the south and central Gulf Coast, grew significantly during the sixteen-year period, despite having been ravaged by multiple hurricanes. After suffering through the 1990s, Homestead grew 100 percent; immediately adjacent Florida City grew 63 percent; Punta Gorda increased by 19 percent; and Charlotte County, which includes Punta Gorda (its only incorporated municipality), grew 11 percent.

But Gulf Coast and Florida Panhandle states are still reeling from the 2004 hurricanes. Biloxi and Gulfport grew from 1990 to 2000, but understandably lost population from 2000 to 2006. New Orleans lost a few thousand head count from 1990 to 2000 but lost a whopping 54 percent after the 2005 hurricanes. Pensacola, Florida, also lost population between 1990 and 2006, during which time the city was struck by three hurricanes. Each time, residents were unable to rebuild before the next one struck. We think of Pensacola as a special case; the likelihood is that it will grow substantially within the next decade unless it experiences one or more additional hurricanes.

We have not conducted extensive analyses of population change in communities that experience significant disasters. On the basis of the work we have done, however, we are inclined to suggest that the major regional and sectoral forces that drive population change will dominate the population trajectories of communities, even though they experience a disaster. But disasters with systemic consequences for the community almost always create significant, although perhaps temporary, variations in the long-term trajectory on the population.

POPULATION CHARACTERISTICS

The city of Homestead became a lot less middle class after the hurricane of 1992. As we said earlier, a large proportion of the total population left. When the U.S. Department of Housing and Urban Development provided resources to build apartments, it provided rental assistance vouchers. Since so many of the pre-event residents did not return, advertisements for the assisted housing drew many poor from other parts of Dade County. Consequently, in the years that followed, the population that left was replaced by people with fewer job skills and far less income. A decade after the hurricane, when Miami spread south to the northern border of Homestead, almost 10,000 homes were soon built near highway and tollway interchanges, and once again the demographic composition of Homestead changed.

In Northridge, California, the 20,000 mostly middle-aged, middle-class people who left after the 1994 earthquake were replaced in the first couple of years primarily by recently arrived immigrants from Mexico and Central and South America, as well as by Korean immigrants who retained much of their ethnic identity. With the change in population, small businesses changed, too. Many small businesses owned by Americans of European descent were replaced by other small businesses owned by Latinos and Koreans. The community changed irrevocably, even though it looks remarkably similar to the pre-event community.

Before the 1999 flood that almost completely inundated Princeville, North Carolina, Princeville housed about 2,000 people. A year later, the population was less than 900. By 2006, the population was back to pre-flood levels, but most of the people there were not the same people who lived there before the flood. Much of Princeville's housing was replaced by public housing projects that attracted people from other areas and changed the characteristics of the community rather dramatically.

In New Orleans, various data sources point to the likelihood that the city's demographics have been dramatically altered since Katrina. Prior to Katrina, about two-thirds of the city's population was black, whites comprised about 28 percent of the population, and Hispanics and Asians accounted for most of the remainder. Now, the population of blacks seems to have declined radically, as evidenced in a recent election in which the majority of voters were white and elections for the city council resulted in a white majority for the first time in over two decades. In general, it appears that New Orleans is becoming a "smaller, whiter city with a much reduced black majority."[17] The city is also seeing an increased number of Hispanic residents, many lured by the prospect of construction jobs.

Some communities experience very little change in community demographics. Los Alamos, New Mexico, lost about 5 percent of its housing units in the 2000 fire, but the dominant employer, the Los Alamos National Laboratory, survived and continued operations despite having lost a number of build-ings. As a result, most of the people in the 400 destroyed houses did not leave; instead, they found other places to live in or near Los Alamos—perhaps because their employment remained stable. Similarly, Grand Forks, North Dakota, retained its general population numbers and characteristics.

LONG-TERM CHANGES

Significant long-term demographic changes in a community seem to depend on several variables, including community economics; the availability of hous-ing, health care, schools, and employment; and the mobility of the existing pop-ulation. The most important appears to be the economy of the community, both pre- and post-event. If the economic "magnet" has not been weakened, people are likely to stay or return as they are able, depending on available housing, jobs, and so on; if it has been weakened, people are likely to drift away per-manently. Before generalizations can comfortably be made about what is likely to happen under various conditions, much more research needs to be done on the demographic consequences of disaster. Nonetheless, local officials should be aware of the range of possible outcomes in their communities following an extreme event, and they should be alert to the likelihood that what occurs after the event will depend to a considerable extent on what happened before it.

RIPPLE EFFECTS

State and federal disaster policies rarely recognize the impacts of disasters on communities that are not directly affected. But the flooding in New Orleans from Hurricane Katrina provides ample evidence of the ripple effects of new populations in neighboring communities. In Jefferson Parish, for example, housing has become scarce and more expensive since the hurricane and flood in Orleans Parish, and serious crime has increased significantly. The hospitals are crowded with patients who have no insurance and cannot pay for the ser-vices provided—patients who previously might have gone to "Big Charity," the public hospital that closed in the wake of Katrina. The state of Louisiana has provided little money to the hospitals that have provided additional ser-vices since Katrina, so the hospitals in Jefferson Parish have been operating at a loss with little prospect of turning that around in the near future. At the same time, one of the hospitals located on the line between Jefferson and Orleans

Parishes has benefited in that it has purchased and reopened several Orleans Parish hospitals that were closed after Katrina. For this hospital, Katrina has produced the positive ripple effect of increased market share and, perhaps eventually, increased revenues and margin.

CHAPTER 8

HOUSING AND REBUILDING ISSUES

"Not going home is already like death."

E. Catherine Tobler, Vanishing Act

Along the Mississippi coastline, steps and slabs were all that remained after the storm surge from Hurricane Katrina cleared a path extending several blocks back from the beach. Photograph taken in March 2007 near Pass Christian, Mississippi.

*T*HE NATURE AND EXTENT OF housing problems following extreme events vary considerably among communities. This chapter focuses on what can happen in a community when a significant proportion of homes are damaged or destroyed.

THE NEED

In every community we studied, temporary housing was needed for people who lost their housing and did not evacuate to undamaged areas. Emergency response and relief teams, including hospital staff, public safety officers, local government officials, and utility crews, also need temporary shelter. Some of these workers will have homes that have been damaged or destroyed, while others will have come from elsewhere. Cleanup crews that arrive from outside the community sometimes bring their housing with them in the form of trailers, campers, and recreational vehicles, but they still need water and sanitary hookups or facilities. After the cleanup crews come the construction crews—usually single people or people traveling alone who are from out of town; because they plan to be there only temporarily, they look for simple, inexpensive housing, such as motels and hotels. Housing is also needed for those who provide services to all these workers: the restaurant workers, hotel staff, and gas station operators. And at some point, temporary housing is needed for returning workers and their families.

In every community we studied, providing the needed housing raised a series of issues as emergency shelter was replaced with temporary shelter, which eventually gave way to more or less permanent housing.

EMERGENCY AND TEMPORARY HOUSING

The horror stories from New Orleans about emergency shelter in the Super Dome and the convention center are atypical experiences in the United States, at least judging from the communities we studied. Our communities did not rely much on tents for emergency shelter; more often, local schools and YMCAs became temporary shelters for a few days or, at most, a few weeks. People who were unable to return to their homes often found places to stay with friends and relatives or in vacant, available rental apartments or houses. Some families left town to find housing, especially if the primary breadwinner had become temporarily unemployed. Before long, FEMA trailers arrived and FEMA trailer parks were created. Or homeowners were able to obtain trailers or recreational vehicles with FEMA assistance, and park them on their lots.

FEMA trailer parks usually begin to take shape within a week or so after the dust has settled or the water has receded. As necessary as they are, however, they are not without issues. Some local officials reported to us that they believe FEMA has a tendency to locate the parks in the "boonies," where there are no grocery stores, no gas stations, no parks, no public transit, and no access to school for children. From FEMA's perspective, those sites probably make sense because it is often difficult to find a site that is sufficiently flat enough, large enough, and vacant enough to put a trailer park. The urgent need to get the facility up and available should presumably be traded off against the need for convenience, but there almost always seems to be some friction between local and federal officials concerning the location of the trailer parks.

We also heard from local officials about conflicts over which organization had primary responsibility for policing the trailer parks. As described earlier, residents already suffering stress from the storm or earthquake often find themselves in cramped quarters living next to people who are very different from them. Clashes arise and, over time, can become frequent and intense. If the trailer parks are located outside the city, police services may be thought to be the responsibility of the county or parish sheriff. Sometimes, however, FEMA officials on site have maintained that they are responsible for policing the park. To the best of our knowledge, the issue remains unresolved, worked out from state to state and disaster to disaster.

DEMANDS ON LOCAL GOVERNMENT

Disasters generate an enormous workload for local government. Some of that workload is likely to be unfamiliar: removing and disposing of tons of debris, reestablishing demolished utilities, and helping to set up emergency housing. But much of the added workload is generated by a greatly increased demand for services with which the local government is familiar, and while there are already policies and procedures in place, those policies and procedures will be severely strained in the aftermath of the disaster.

Initial Inspections

Local officials reported that their building and health departments had massive workloads for months and even years following an extreme event. In most communities, the first set of demands grew out of the need for prompt inspections.

In the case of earthquakes, inspections were needed to ascertain whether buildings could be occupied. Green tags were given for structures that were available for immediate occupancy, yellow tags for buildings requiring repair, and red tags for buildings that were unsafe for occupancy. The need for the inspections and inspectors is well-known in California, so there are arrangements for large numbers of inspectors from other jurisdictions to supplement the local staff.

Flooded areas and places struck by tornadoes and hurricanes usually require building inspections, not only to ensure that the structures are safe for occupancy but also, when warranted, to search for bodies. When the number of structures is large, the demand invariably exceeds the number of staff members available from the local governments. In those cases, inspectors must be brought in from elsewhere, and they almost always need training to become familiar with local code variations and idiosyncrasies.

Contractor Licensing

Most local jurisdictions require that contractors be licensed to do business in that jurisdiction. After an extreme event, the demand for contractors often greatly exceeds the number licensed in the jurisdiction. Tensions arise when building owners are eager to begin repairs and the contractors they want to use are not licensed in the community. However, local building departments are reluctant to waive the licensing requirement because they want to guard against fly-by-night contractors who flock to disaster sites hoping for work.

Local jurisdictions rarely work out policies for handling licensing procedures under emergency conditions in advance; they usually wait until after the disaster happens and unhappy people are lined up at city hall complaining to council members and the mayor or chief administrative officer. By this time, there is an almost endless list of other issues to deal with, staff members are already fatigued and tired of sleeping in city hall, and there is no time to work out a policy that has been thought through carefully.

Permits and Construction Inspections

In the aftermath of a disaster, local building officials are almost always besieged by property owners wanting permits to repair and improve their property. In virtually every community we studied, the demand for permits and construction inspections overwhelmed local staff. At the same time that other local officials are beginning to get some relief from the emergency burdens imposed by the disaster, the workload for building department staff is usually increasing. It seemed to us that turnover rates in building departments were particularly high during this period as staff members either burned out or were considered by higher-level officials to be not quite up to the job.

A few local governments we studied established priorities for inspections. Facilities that were needed in the community got priority inspections by local and, as applicable, state inspectors. These facilities included restaurants, hotels and motels, and hospitals, health care facilities, and other critical or chronic care facilities. Setting priorities seems almost essential to facilitate rebuilding and recovery.

Requests for Exemptions, Waivers, and Variances

In virtually every community we studied we found that the number of requests for exemptions and waivers from zoning and building regulations increased significantly after an extreme event. Perhaps the most common waiver requests involved requirements that were triggered when repairs or improvements exceeded some fixed percentage of the building's value. For example, many jurisdictions require that if repairs or improvements exceed 25 percent (or some other percentage) of the building's value, the entire building must be brought up to current building and specialty code standards. That may entail meeting the requirements of the Americans with Disabilities Act and making electrical and plumbing changes. Because building owners want to get their buildings back in use quickly, they may pressure local officials to waive those requirements or, more subtly, to estimate damage at less than the specified percentage. Requests may also be made to waive contractor licensing requirements; zoning, setback, and side yard requirements; and virtually any other regulation that stands in the way of a quick repair. Local officials reported being torn between the need to get buildings back into use quickly and the desire not to recreate the structural vulnerabilities that contributed to losses in the first place.

In some places, rebuilding may be substantially delayed while various stakeholders wrangle over proposed changes to land use regulations—changes that often have material implications for homeowners and small-business owners who are compelled to wait until the dust clears, lest they find out that their renovation or rebuilding does not meet some altered code or regulation. An example from New Orleans pertains to the requirements to raise homes above some identified height. As described in a *Washington Post* article,

> "Substantially damaged" houses in the area now must be raised, often three feet above the ground. But the requirements contain enormous loopholes, and there is a huge financial incentive to avoid them.

Raising a house can cost upwards of $50,000, especially for the modern suburban homes built on concrete slabs in some of the most flooded areas. The federal government offers grants of as much as $30,000 for repairs, but in many cases much more is required.

"The vast majority simply do not have the financial resources to rebuild differently," said Greg Rigamer, chief executive of GCR & Associates and a consultant in the rebuilding.

Residents could avoid having to comply with the new guidelines by getting permits before the rules were enacted locally—thousands in New Orleans did—or if their houses were determined to be less than 50 percent damaged by Katrina.

Many homes, even those that took on 10 feet of water for weeks, have been designated beneath that threshold, including hundreds whose owners appealed larger initial damage assessments.[18]

Other common requests for variances come from homeowners seeking extensions to the length of time they can keep their temporary homes parked on their lots. It is not uncommon for the vehicles to remain on homeowner lots for years, particularly when the homeowner was uninsured or underinsured, or when the demand for contractors far exceeded the supply. Local officials have to make difficult decisions with the homeowner on one side and vocal, unhappy neighbors on the other.

HOUSING COSTS AND LAND VALUES

What happens to housing costs and land values depends, of course, on the interplay of supply and demand. That interaction played out a little differently in every community we studied and appears to be affected by several factors. In some Gulf Coast areas with severe damage at the shoreline, land values escalated rapidly and almost immediately after Hurricane Katrina. As property and casualty insurance rates skyrocketed during 2006 and 2007, the escalation apparently slowed somewhat—at least temporarily. Almost everywhere at the water's edge, smaller homes that were damaged or swept away are being replaced by much larger homes, and small buildings with rental units are being replaced by larger hotels and condominiums or by commercial buildings catering to shoreline business.

In other local jurisdictions, housing prices appear to have increased in proportion to the amount of housing that was damaged or destroyed. However, land prices do not seem to have increased like they have along the Gulf Coast shoreline.

Other variables include the availability of housing in a nearby surrounding area, the number of residents who left the community in the wake of the event, and the extent of activity by speculators bent on acquiring special land, such as beachfront property.

CHAPTER 9

SOCIAL AND PSYCHOLOGICAL CONSEQUENCES

"It's a never-ending nightmare. It never goes away."

A Northridge earthquake survivor

Humor is a well-documented means of dealing with psychological stress. Photograph taken in Lakeview neighborhood of New Orleans in March 2007.

*P*ERHAPS THE MOST OVERLOOKED AND understudied part of disasters has to do with individual, family, and community social and psychological well-being. Following the initial shock of injury or death to friends and family and of damage to the built environment, people try to make sense of what happened and find ways to deal with it. We have grouped our observations in communities across the country into several broad categories: (1) expectations about the aftermath; (2) individual mental health issues; (3) community healing; and (4) cohesiveness and divisiveness.

EXPECTATIONS

The social and political consequences of disaster are not easily predicted, but they seem to follow a general pattern. In almost all communities we studied, as the immediate emergencies have been addressed and people are cleaning up and making repairs, there seems to be a generally expressed optimism that "this is just a bump in the road, and we'll soon be back where we were before it happened." We have not heard this from everyone, but we have heard it from most local elected officials and many residents, even in places that suffered almost total devastation. We could not tell how much of what we heard was rhetoric born of belief, heartfelt hope, or sheer desperation, but it suggested an expectation that things in the future could be much like they were before the event. It may be that no one wants to appear on television glumly predicting, "We will never be able to recover."

In the cleanup, fix-up period early in the aftermath, expectations about a return to normalcy are typically bolstered by cooperation and mutual assistance. Neighbors help neighbors. We often heard, "We've never seen anything like it. This disaster brought everyone together." This sense of community is further enhanced by a swell of volunteers who come to the community to give comfort and to help with clearing debris and rebuilding. But except in New Orleans, where volunteers continued to arrive three years after the flood, volunteers usually begin to drift away after a few weeks or months, and residents may feel left largely to themselves.

Later in the aftermath, individuals start to define, usually quite subconsciously, a "new normal" to replace what they had defined as normal before the event. At the same time, individuals, organizations, and groups find that their concerns and agendas are not shared universally in the community. Other people have other interests, which sometimes conflict. Pre-event agendas for the city council and nongovernmental civic groups go by the board as the meetings are dominated by post-disaster problems and concerns. Traditional processes may be short-circuited. People who were rarely or never involved in political exchange join the fray in order to get their issue addressed.

Early in the aftermath, community and local government expectations about state and federal assistance are sometimes unrealistic. Funds are typically project based and designated for specific uses. Most of the grants are for brick-and-mortar projects. The federal government does not provide grants to cover local government operating expenses; it provides loans for that purpose. Moreover, federal programs are not particularly flexible; one size is usually expected to fit all, and every dollar spent must be accounted for.

As issues in the community move to the governmental agenda, unrealistic expectations are sometime placed on local officials. For these local professionals, who are dealing with problems they have never before faced, added stress comes from elected officials and community groups who want the problems solved NOW. In some communities, this results in frequent turnover in key positions. We do not want to reference specific communities, but in some that we visited, the positions of city manager, chief financial officer, and building official turned over almost annually for several years after the disaster. The combination of long days and weeks, political pressure, and no time for personal lives gets to be too much. People leave of their own accord or get fired.

After a few weeks, residents who had been focused on the losses to the built environment and on injuries to people they know begin to sense the losses to the social, economic, and political community, although few ever put it into those words. Expectations begin to change. The "feel good" period of cooperation and support starts to fade. People in every community get discouraged. In New Orleans, for example, virtually every locally generated evening newscast, even two years after the flood, contained flood-related stories and not many of them were particularly pleasant. Add to that the plight of so many people who are making mortgage payments on a house they can't live in, as well as payments on the home somewhere else they established after the flood. Living in a small trailer parked on your driveway for a week is an adventure. For eighteen months or two years or more, it is a continual reminder of one's plight. Faced with daily reminders of the extreme event and dwindling attention from outside the area, people begin to wonder if they have been forgotten.

INDIVIDUAL MENTAL HEALTH

In our research, we did not plan to look at issues of post-disaster mental health. But when we asked people, "What happened?" many made it abundantly clear that long-term psychological stress and, often, clinical depression were serious problems for them. The stress and depression made it difficult for some to recover personally, even after a decade or more. These problems rose to the community level as psychiatric services were stretched to the breaking point. Interviews with hospital personnel in New Orleans, for example, were routinely punctuated with concerns about the relative dearth of available psychological services. "What we really need, right now," they said, "are more psych beds. These folks are ending up in our emergency rooms, unable to get the care they need and complicating the care available for others."

Both men and women experienced the emotional stress and psychological problems. We heard about marriages that ended within a year or so of the disaster as a direct result of behaviors by one or the other partner in the weeks and months following the disaster. In most communities, we interviewed people who said that they were being treated for psychological ailments, disaster-related depression, and anxiety even five years after the event; some respondents reported having been in treatment for six or seven years post-event, and some had attempted suicide. Many times, people we were interviewing broke down in tears, even though the actual disaster was five or six years in the past. It was painful to interview them, knowing that there was little or nothing that we could do, other than listen.

An analysis of our interview notes suggests that long-term adverse psychological effects are related to an individual's and family's resiliency and to the extent of their losses and the disruption to their "life trajectories." A few of the local officials we interviewed reported that the local government had made counseling services available for "a month or two" after the disaster but that very few people took advantage of them. Unfortunately, in many cases, the problems do not develop or emerge until later, when little help is available. Post-traumatic stress does not follow government-imposed timelines.

We have concluded that depression and other psychological illnesses likely exacerbate the challenges associated with recovery. After all, a person who doesn't want to get out of bed probably doesn't care about reclaiming his or her home.

One of the most emotionally difficult parts of our research was listening to parents tell us about the disaster effects on their children. Some people reported that they and their children had difficulty sleeping in the months following the disaster, and that the children were particularly hypersensitive to unexpected noises or weather changes for as long as five years afterwards. In one community devastated by a tornado, a teacher told us, "For the past three years or so, whenever the children look up and see low-hanging, dark clouds forming, they stop whatever they are doing and run for home." One woman told us about her grandson who, as an infant, was sucked from his second-story bedroom by a tornado. He suffered brain damage from his fall, and for the entire family, every day is a reminder of the disaster. Several years after a tornado hit a community, we were interviewing a local official and were interrupted: "Please wait," he said. "This is my son on the phone. He is at the community swimming pool and it looks stormy. He wants to come home right now, and I'm going to go pick him up." We waited.

We heard the same thing from parents in earthquake country. "Ever since the earthquake, every time there is even a little shaking or there is a strange noise, the children get wide-eyed and look for security." Another woman told us that her children could not sleep through the night for months and months after the earthquake, and that the trauma contributed to the end of her marriage. In Tarboro, North Carolina, a nonprofit group from the community arranged a way for children to talk through their fears and concerns stemming from the flood that inundated their city. The group contracted with an organization from Colquitt, Georgia, called the Swamp Gravy Institute. An art service organization and the consulting and training arm of the Colquitt/Miller County Arts Council, the institute held sessions with children in Tarboro to have them share and role-play their experiences during and after the flood in an attempt to facilitate their ability to cope.

FEMA and the Department of Health and Human Services provide funds to support assistance and counseling to individuals experiencing mental health problems caused or aggravated by an extreme event or its aftermath once a Presidential Disaster Declaration has been issued.[19] When available, federal assistance for counseling is provided through state government, but the state first has to demonstrate that the demands caused by the extreme event outstrip its capacity to provide service, and it must apply for the funds within fourteen days of the disaster declaration. The program provides for up to nine months of service. However, not all disasters are issued a Presidential Disaster Declaration, and for those that are not, a local government must look to its state or use its own resources to help people deal with disaster-induced mental health problems.

The need for counseling and mental health care, in our experience, may not even be evident until much later than fourteen days after a Presidential Disaster Declaration and may go well beyond nine months. We believe that it would be prudent for local government officials, perhaps through their state-wide associations, to work with state mental health officials to develop a care policy and funding provisions before the next disaster, or to seek a change in federal policy.

COMMUNITY HEALING

Community catharsis is necessary. Virtually every community we visited conducted a memorial on the first anniversary of the disaster. Most memorial events include prayers for those killed or injured and for a brighter future. Permanent memorials take different forms: some communities erect a plaque in memory of those who died or simply in memory of the event. Others create a monument, a park, a statue, a facility, or some other brick-and-mortar edifice. Grand Forks, North Dakota, erected signs on vacant lots indicating what used to be there. Gulfport, Mississippi, created a quiet place in a park near the beachfront and a monument containing items found in the rubble from the retreating storm surge. In the Internet age, many communities have established Web sites and blogs commemorating their experiences with photographs and text. The photograph and blog spots established in the wake of Katrina, for example, number in the thousands and catalog the hurricane's devastating effects on Louisiana and Mississippi.

COHESIVENESS AND DIVISIVENESS

We found that while the length of the coming-together, "feel good about ourselves" period varied by community, the period tended to be one that people remembered fondly for years after the event. In nearly all cases, however, it dissipated before the cleanup was complete. Early in the aftermath of a disaster, the community shifts from a period of collaboration and cooperation to a period of political conflict about who gets what and when. In some cases, the conflict is exacerbated by preexisting community tensions. Equity and fairness issues almost invariably arise among those who suffered losses and those who didn't; those with losses sometimes group themselves in terms of who lost the most. Along the Mississippi Gulf Coast, in the wake of Katrina, some people referred to themselves as "slabbers"—people whose homes and belongings were swept away completely by the storm surge, leaving only the foundation slabs. In some communities, an "us vs. them" mentality may develop among victims toward "nonvictims"—those who did not lose as much and who may be receiving more in the way of assistance than others think they are entitled to.

Except in New Orleans, we did not see or hear of anyone overtly playing "the race card" in a public forum. Nor did we find any systematic policies or practices that directed benefits primarily to one racial or ethnic group at the expense of others. In some cities with river flooding, the poor suffered most of the damage because they lived in less desirable lowland areas. In cities where riverfront property was desirable, those who could afford riverfront property lost the most.

PART 3

POST-DISASTER EXPERIENCES: WHAT HAPPENED TO LOCAL GOVERNMENTS

IKE ANY OTHER ORGANIZATION, A local government often experiences losses from extreme events: municipal buildings and infrastructure are damaged, financial arrangements are disrupted, data are often lost or destroyed, and employees may not return to work. For a local government to help the community recover, it must itself recover sufficiently to be an effective agent. In this section, we identify the major effects that we saw on local governments across the country, whatever the cause of the disaster.

CHAPTER 10

LOCAL GOVERNMENT WORKLOAD AND EMPLOYEE STRESS

> "I haven't been home except to shower and change clothes for more than six weeks. I sleep, eat, and shave here."
>
> Midlevel municipal manager in City Hall

EXTREME EVENTS TRIGGER EXCEPTIONAL WORKLOADS for public employees. These heavy workloads, which are periods of high stress, begin before the extreme event occurs—at least in the case of floods and other events that one can see coming—and extend through emergency response, initial recovery efforts, and long-term recovery efforts.

Anyone who has had experience in the construction of local government facilities understands that building, repairing, and restoring water treatment and distribution centers, wastewater collection and treatment centers, schools, hospitals, bridges, and other facilities are time-consuming, complex, and difficult tasks. Even when strategies for implementing complex, large projects— such as design-build contracting—are used, big projects typically require years to move from preliminary design through permitting, construction design, contracting, site approval, and the like before construction can begin. Imagine a local government having to repair or build anew multiple facilities all at once following severe damage from an extreme event. The workload can be truly staggering, even for the most effective local government, especially when the need is urgent and when other communities in the same situation are competing for available contractors and labor. If the disaster is regional, building supplies and materials, labor, and managerial talent will be in short supply and will increase in costs—often dramatically.

Add to this the homeowners and business owners who want to repair, rebuild, and reinhabit or reopen quickly, and who need permits, inspections, and licenses as well as an enormous amount of information on any number of issues. Then add the massive load of internal paperwork and accounting needed to apply for state and federal disaster assistance. This workload, which is already overwhelming, is then often compounded by the need to work in offices that have been damaged or destroyed, or to make do with temporary quarters that are never quite adequate, that do not have needed files available, and that have temporary computer systems in place through which needed information is not easily accessible.

It isn't pretty. It isn't fun. It is hard, demanding, and exhausting work. Most of us haven't been trained to work under these stressful conditions, and we haven't had much experience dealing with the enhanced variety of challenges. Yet time and again, we saw local officials facing these very challenges and doing their best to serve the citizens who need them.

WORK OVERLOAD

The workload doesn't diminish for local government after the immediate need to put out fires, rescue those trapped, recover bodies, and provide temporary emergency services to survivors. It just shifts to other arenas. Even before the debris is removed, local government finance, planners, and public works staff face the mountains of paperwork that are required to apply and account for federal financial assistance. Project worksheets must be completed, data that are not readily available must be found, complex financial arrangements must be made quickly, and plans for housing and other needs of the community must be developed quickly.

At the same time, as noted above, the staff is handling increased demands for permits and inspections (following Katrina, for example, demands for building permits in Biloxi increased fivefold almost overnight—from 1,000–1,500 in each fiscal year (FY) between 1999 and 2005 to 5,000 in fiscal year 2005–2006),[20] arranging for the repair of public infrastructure, managing clean-up and construction projects, and responding to calls from residents on a host of problems. Local officials find themselves dealing with two sets of demands at the same time: the routine local government functions that continue and the new set of disaster-related activities.

The demands are intense, and they continue for months without letup. After a while, the adrenaline is gone, but employees must labor on. Dealing with people from state and federal granting agencies is not easy; in the case of large disasters, the federal government brings in temporary help who are not always well-informed. We often heard local officials say, "Ask the same question of five of them and get five different answers," or some version of that statement. New staff members rotate into the community, requiring the added load of starting up with a new staff member. Rules and regulations for various federal programs are often complex and change almost continually. Frustration often builds among the local, state, and federal officials, sometimes boiling over into outright conflict.

UNMET EXPECTATIONS AND NEW ROLES

In most cases, elected officials and department heads are typically just as inexperienced in disaster management as are lower-level local government employees. Lack of experience necessitates "on the job" learning, however difficult that may be. Role conflicts are likely to arise as department heads feel that their turf is being infringed upon by others. Accustomed to working in "solo" environments to administer their programs, these managers often find themselves being asked to work cooperatively with others to solve complex problems. Doing so requires learning how to communicate with and appreciate the skills of and the values held by different departments. This is no easy task in good times; it is even more difficult under pressure.

Sometimes decision makers may feel that progress isn't being made quickly enough or that the outcomes of actions are not entirely satisfactory. Elected officials, feeling pressure from constituents, may have unreasonable expectations about the pace of recovery and about what local government staff is able to do about it. To meet the public's demand for accountability, they may make fundamental attribution errors, assuming that others are personally responsible for any mistakes that are made, discounting the role played by external factors such as simple misfortune. They may also fall prey to self-serving bias, taking responsibility for their own successes while attempting to deflect blame for their own failures. However, neither pointing fingers nor taking credit for early successes helps the process of recovery; rather, each response focuses attention away from needed problem solving.

CONFLICTING DEMANDS BETWEEN HOME AND WORK

In addition to the increased workload and the new problems that emerge daily, many public employees must confront the personal, inner conflict between the need to help strangers and the pressure to help their own families at home. These employees are expected to remain at their posts as needed during emergencies. Some—emergency managers, sworn public safety officers, utility workers, medical personnel, and chief executives—know that they will be required to stay and understand what is expected of them. Others—for example, buildings and grounds staff, building officials, financial personnel—are less likely to expect to remain at work but may nevertheless find themselves called upon to stay or to report for duty as the storm winds up or winds down.

Yet like the citizens they serve, these employees are also very likely to have homes that have been damaged or destroyed and families who have likely been adversely affected. Depending on whether there was any warning of the extreme event, emergency and relief workers may not have had an opportunity to communicate with their family members to learn where they are, whether they're prepared, and what they might need. For employees who have not consciously signed on for such duty, it is difficult to literally work, eat, and sleep at city hall for days or weeks, and to come home, if indeed there is a home, for only a change of clothing and a quick hello and goodbye. Nor is it easy for their families.

In virtually every city we visited and studied, public employees stayed at the job, eating and sleeping at their workplace for extended periods while their families and their homes had to wait until everyone else was taken care of. They did so without thanks from the community and, often, without extra pay or, almost always, without bonuses. In New Orleans, there is evidence that some public employees left their duty stations without authorization. That behavior by public officials is reprehensible, but it is also extraordinarily rare in most jurisdictions. Most public officials—and their families—sacrifice beyond the basic call of duty in the event of a disaster.

CONSEQUENCES OF EMPLOYEE STRESS

In many of the cities we studied, we were told, "Oh, he (she) doesn't work here anymore. Left a couple of weeks ago." Local government employee turnover in the two or three years following an extreme event was significantly higher than we had originally expected. City managers, finance officers, and building officials seem to be particularly vulnerable. The pressure that local elected officials experience regarding the pace of recovery and the problems that are still unsolved is often passed on to government employees, who are doing their best against almost insurmountable obstacles. Frustration mounts and greener fields beckon.

Worker performance often declines in the face of extreme and prolonged stress. Employees faced with the same, seemingly intractable problems day after day experience burnout and look for an escape. Sometimes employees are fired, but many more leave their job for a place where they think people will appreciate their work and where the stress will be more manageable. Simply put, the skills used to perform local government jobs transfer to other fields of employment, many of which do not require people to deal with hordes of dissatisfied community members or visible reminders of their own devastated lives.

We think another possible reason for high turnover is that disasters sometimes generate a high demand for scapegoats, and there is only a limited supply of targets. Elected officials are often better at finding fault than they are at finding solutions, so it is far better to blame federal, state, and local government workers for not meeting expectations than it is to blame the public—or themselves, for that matter—for voting down proposals to spend money that would have been used to mitigate the natural hazard before it occurred.

CHAPTER 11

A DIMINISHED REVENUE BASE

ALONG WITH ITS FACILITIES AND infrastructure, a local government's revenue base is often devastated by an extreme event. A local governments is, of course, deeply concerned about developing and maintaining a tax base to generate the revenue that enables it to do what needs to be done. Decades ago, local governments generally depended primarily on revenues from the *ad valorem* ("according to the value") property tax base. Although today they have a more diversified stream of income on which to rely, much of that stream may be reduced or eliminated, at least for a time, following an extreme event. In this chapter, we explore the sometimes surprising impacts that extreme events may have on various local government revenue sources. Since local government taxation practices vary widely among the states, the discussion is kept at the general level except for specific illustrations.

THE REVENUE BASE AS A COMBINATION OF SOURCES

It isn't so much that one element of the tax base is diminished. It is that the extreme event has the potential to diminish most of the local government's sources of revenue. The individual tax bases are linked to one another. Damaged real property is likely to lead to lower revenues from property taxes, sales taxes, tourism, and user fees.

For example, Biloxi's local government relies heavily on three sources of revenue: property, sales, and gaming taxes. The gaming taxes come from a number of casinos located in the city. As shown in Figure 3, between 1992 and 2005, Biloxi's revenue increased more than 225 percent. But because some of the city's casinos

were heavily damaged by Hurricanes Katrina and Rita in late 2005, locally generated revenue to the community fell dramatically after those hurricanes: in a
single year, from 2005 to 2006, total revenue from all three sources dropped by a
fourth. As structures are replaced, the sandy beaches are cleaned, and the casinos
are reopened, tourism will return and the revenues will presumably increase to
previous levels. The question remains as to when that will occur.

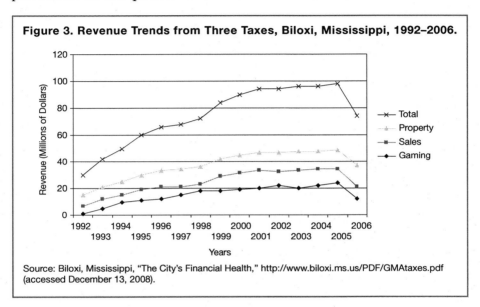

Figure 3. Revenue Trends from Three Taxes, Biloxi, Mississippi, 1992–2006.

Source: Biloxi, Mississippi, "The City's Financial Health," http://www.biloxi.ms.us/PDF/GMAtaxes.pdf
(accessed December 13, 2008).

Sometimes the drive for enhancing local government revenue conflicts with
other important community goals. In Escambia County (Pensacola), Florida,
for example, the need for revenue runs counter to the desire to make the community safer against extreme events. The county is relatively poor except for
areas near the shore: it owns much of Santa Rosa Island and its beautiful white
sand beach on the Gulf of Mexico. A large proportion of the island is within
the federally owned Gulf Islands National Seashore; however, the Escambia
County owns Pensacola Beach, which comprises much of the land outside
the park that is available for development, and it leases that land to people
and corporations that build on it. Because it is legally prohibited from levy-
ing taxes on real property located on Pensacola Beach, the county depends
on head, bed, and sales taxes collected there to shore up its revenue needs.
Its revenue stream would benefit greatly if the older single-family homes and
rental units destroyed by recent hurricanes were replaced by larger hotels and

resorts, and despite the reality of occasional devastating hurricanes, developers want to increase density along the white sand beaches. Thus, we expect
that Pensacola Beach will be much more densely developed over the next few
years than it was before the string of hurricanes that leveled so much of it,
because of the need for local government revenue and the desires of profit-
minded developers.

AD VALOREM PROPERTY TAX

Local governments typically levy *ad valorem* property taxes on real property, personal property, and business personal property. Real property refers
to land, including not only the face of the earth but everything of a permanent
nature over or under it, such as structures, minerals, and timber.

When Homestead was struck by Hurricane Andrew in August 1992, the
city and the surrounding area suffered massive property damage. Direct damage from the winds and water was estimated at $25 billion.

> Andrew's impact on southern Dade County, Florida was extreme
> from the Kendall district southward through Homestead and
> Florida City, to near Key Largo....Andrew reportedly destroyed
> 25,524 homes and damaged 101,241 others. The Dade County Grand
> Jury reported that ninety percent of all mobile homes in South Dade
> County were totally destroyed. In Homestead, more than 99% (1167
> of 1176) of all mobile homes were completely destroyed. The Miami
> Herald reported $0.5 billion in losses to boats in southeast Florida.[21]

Not all extreme events wreak damage that is so extensive, but many result
in damage to a major proportion of the community. In St. Peter, Minnesota,
for example, a tornado damaged or destroyed 44 percent of the buildings,
including every building on the campus of Augustus Adolphus College.
In Montezuma, Georgia, the entire central business district was flooded for
days. When that much damage is done to real property in a community, it
will undoubtedly be reflected in the assessed valuation of real property in
the local jurisdiction. And whereas the impact on the property is immediate,
the impact on local government tax revenue may not be. Most jurisdictions
base their *ad valorem* real property tax for the current year on the value of the

property assessed the previous January. Thus, there is typically a one-year lag between when the taxes are assessed and when they are due. That means that tax collections in the first year following the disaster usually do not reflect the loss to the tax base, particularly if the taxes have been accumulated in escrow by mortgage holders. If the taxes have not been held in escrow, many hard-hit property owners may be unable to make the tax payment. After several years of nonpayment, the municipal or county jurisdiction will be able to foreclose on the property, but that may not be desirable in the quest for recovery.

Local governments take various approaches to valuing property following an extreme event. Communities that have been flooded may be a little slow to reassess property where structures have not been washed away or completely destroyed, so assessed property values remain fairly static until repairs are completed. In jurisdictions where the real property has burned to the ground, blown away, or collapsed from earth movement, property values are typically adjusted more quickly. Local tax boards of appeal are inundated with complaints if they do not make tax adjustments for properties that are no longer there as of the January 1 assessment date.

There are a few anomalies. In some Florida Gulf Coast communities, prices for shoreside and near-shore property increased rapidly immediately after devastating hurricanes. By demolishing and blowing away older, smaller buildings and other low-density uses, the hurricanes effectively cleared the way for larger and more expensive structures, including condominium projects and hotels. In one location we found small residential lots along the shore where only the concrete slab remained; these lots were up for sale with a listed price of $2.5 million.

Personal property consists of such things as furniture, fixtures, plant equipment, office equipment, machinery, boats, aircraft, mobile homes, and recreational vehicles. A key characteristic of personal property is that it can be moved without damage to either itself or the real estate to which it is attached. Many states and local governments impose *ad valorem* taxes on personal property, whether it is owned by individuals or businesses. If you traveled the Gulf Coast after Hurricane Katrina, you would have seen how the personal property tax base fared: boats were found upside down two miles from the nearest water, recreational vehicles were crushed and some had been dragged out to sea, and flooded automobiles were abandoned everywhere. A massive proportion of the personal property tax base of state and local governments had been destroyed.

The effects of the loss of the real and personal property tax bases can have a devastating effect on community revenue for years as businesses and individuals work to repair, rebuild, or replace what was damaged.

(USUALLY) CENTRALLY COLLECTED AND (USUALLY) LOCALLY SHARED TAXES

State governments routinely collect some taxes and distribute them to local governments; collection and distribution are either based on where the revenues were generated or based on a formula intended to promote income equity among jurisdictions. Sometimes local governments levy income taxes, but they can rarely collect them; sometimes they are able to enact an "add-on" to a state sales tax and get that money transferred back to them. Some kinds of business personal property are taxed by the state, and the funds are allocated among the local governments in approximation to the amount of business the enterprise does within the local jurisdiction. Revenues from legalized gambling casinos and activities are garnered by the state and allocated among jurisdictions in one or another way. Some enterprising local governments tax gambling in their communities to generate revenue that is, they hope, mostly from tourists.

School aid comes from centrally collected and locally shared tax revenues. It is almost always allocated on the basis of the school census that each U.S. school district conducts at some time specified by each state early in the fall semester. School districts that are hit by disasters and lose a substantial number of students, if only temporarily, can find themselves in desperate financial straits unless the state adjusts the formula to hold them harmless for some period of time. Similarly, if state aid to a general-purpose local government is formula based, the local jurisdiction may find itself in serious jeopardy financially.

To the extent that sales taxes are returned to communities that generate them, the damage caused by an extreme event can have a devastating effect on retail sales within a community for an extended period of time.

TOURIST TAXES

Tourist taxes consist of levies on the kinds of things that tourists need when they visit a community: rooms, restaurant meals, rental cars, and almost everything else that can be conceivably taxed in an attempt to "export" taxes to nonresidents.

But the tourists stop coming when the beach is littered with the debris left by the hurricane, when the hotels and motels are destroyed, and when the casinos are gone. In short, they stop coming when the things that would attract them are no longer there. In the Wisconsin Dells area, after the levee creating Lake Delton failed in June 2008 and the entire lake drained into the Wisconsin River, one resident who depended on tourists for income lamented, "After they see the mud where the lake used to be, then what are they going to do?" For communities that rely heavily on tourism as a primary economic driver, getting the tourist-attracting amenities back in place and advertised is critical to their economic well-being and to the financing of local government.

USER FEES AND UTILITIES CHARGES

User fees include fees for local government utilities; charges for necessary and optional services; charges for the use of government-owned facilities, such as community centers and park shelters; fees for recreation activities; and fees for individuals and businesses for licenses and permits. Such fees have become a major source of income for local governments. In 2004, the Wisconsin Legislative Audit Bureau conducted an extensive analysis of user fees by Wisconsin local governments; it reported that, collectively, local governments in that state charge more than 500 separate user fees and that revenue from those fees in 2001 totaled 21.1 percent of all local government revenue in Wisconsin.[22]

Homestead, Florida, operates its own electric utility, generating, distributing, and selling electricity. When it is economically advantageous to do so, the utility buys power. Prior to Hurricane Andrew, the municipal electrical utility was a major source of revenue for the municipality, but the hurricane winds snapped off virtually every power pole in the city. The utility had employed far too few people to replace the poles and to quickly rewire virtually the entire city. External help was essential, but the task was still enormous. During repairs and the rebuilding of the city—a time when the city needed revenue desperately—revenues from the utility dropped significantly because it was a long time before it was able to deliver electric power again to those buildings that were still standing.

CHAPTER 12

SKYROCKETING EXPENSES, CASH SHORTAGES, AND CLOSING THE REVENUE GAP

"We need the employees, but we had to lay them off.
No money."

Local government official in a devastated community

*A*CHIEF EXECUTIVE TOLD US IT was necessary to lay off a significant number of county workers in the wake of the disaster, just as those workers were needed the most to deal with enormous, recovery-related workloads. Expenses were soaring, revenues were plummeting, and the county simply could not meet the payroll. This county suffered more than most from skyrocketing expenses and major revenue shortfalls, but every jurisdiction we studied had a similar story.

EXPENSES RISE RAPIDLY

Local government must be able to function to help the community recover. The utilities they provide must be restored and operating at some temporarily acceptable level of service and reliability. Schools and government-owned hospitals must be returned to operating condition, staffed, and equipped to carry on, along with city hall and all other critical facilities. Streets and highways must be cleared and reopened; bridges may need to be repaired or replaced. Even if only a portion of the community was damaged, the undamaged areas still require the services they have come to expect from local government—even while the damaged sectors require extraordinary efforts.

Staffing demands for local governments typically increase during the aftermath of a disaster: employees must perform the regular, routine array of duties to undamaged parts of the community at the same time that they're struggling to put the machinery of local government back together and dealing with an increased demand from citizens for services. Homeowners and business owners need permits and building inspections; contractors from outside the city want to be licensed to work in the jurisdiction. All of this extra work costs money for which the local government, most likely, did not budget. Besides overtime pay and temporary employees to meet the extra workload, the extra costs show up as rental costs for temporary working space and equipment lost in the event, material for repairs and reconstruction, and even supplies that may have been lost. And all these disaster-related expenses must be accounted for.

Some expenses, such as those for normal government operations, are not reimbursed. The local jurisdiction can apply for loans to cover those expenses, but the federal government expects to get that money back. If revenue shortfalls are severe for an extended period, the loans are difficult to repay. This is especially a problem for jurisdictions with large numbers of lower-income people, high service demands, and inadequate tax bases.

Some expenses, such as clearing debris from roads designated as state highways, will most likely be paid for by the state government unless the state undertakes those activities itself. Sometimes other local governments, operating under a mutual assistance pact or simply good will, may provide services at little or no immediate cost to the local government in which the disaster occurred. Those governments can invoice FEMA for their costs, as long as the helping jurisdiction has accounted for those costs appropriately.

Even when a local government expects federal reimbursement for disaster-related costs, the service must be paid for before reimbursement can be sought. Thus, local governments invariably find themselves facing a bridge financing problem. Somehow, they have to be able to carry themselves financially until the earmarked check comes in from Washington. And that could take a long time, especially when the money flows through the state government and not directly into the local jurisdiction.

MEETING THE IMMEDIATE NEED FOR CASH

Most local governments do not maintain large balances in their checking accounts. Funds are invested and drawn upon as needed through the fiscal year, so liquidity may not be an especially important indicator of organizational health. Local government finance officers are responsible for ensuring that funds are available for immediate needs, such as payroll and regular expenses, but also for managing cash flow to ensure that as little money as possible sits idle. Thus, even under the best of circumstances, local government officials have to hustle to free up funds to meet immediate needs in the wake of a disaster. Those funds may be obtained through reserves, insurance, loans, and institutional arrangements within states.

Reserves, Insurance, and Loans

Some local governments are fortunate to have significant reserves, although most states do not allow local governments to set aside significant amounts of money "just in case." Reserve funds are almost always earmarked for some specific purpose, such as debt repayment, facilities improvement and replacement, and accrued employee sick leave and vacation. Following disasters, communities with reserves often borrow from those reserves to finance operations and help rebuild necessary facilities. That is, the local jurisdiction borrows from itself, with the expectation that the loan will be repaid either from federal grants (for those expenditures that are eligible for federal reimbursement) or over time as the revenue situation improves. The laws on what can be done with reserves vary from state to state, so local officials should determine what their state allows.

We encountered a few local governments that actually had disaster insurance. Disaster insurance is not always available from commercial sources—at least not at a price that local governments are willing to pay. Consequently, local governments that are insured are usually insured by "pool" policies that are created by a group of local governments themselves. Local governments contribute to the pool, and they draw from the pool when they have losses. Stuart, Florida, for example, was able to pay for a substantial part of its hurricane losses with proceeds from its pool insurance.

"Biloxi hit the jackpot when two months before Katrina the city spent $92,000 to purchase $10 million worth of business interruption insurance to cover lost casino revenues."[23] Biloxi's mayor, A. J. Holloway, is said to have taken out a business interruption insurance policy on the city's casinos because the casinos provided, directly and indirectly, a major share of Biloxi's revenue. The city made only one or two premium payments before Katrina's storm surge struck, putting the casinos out of business for some time. As of July 2006, the city had collected $7.5 million from the insurer, which helped considerably in covering costs incurred from the storm surge and flooding.

Communities sometimes find themselves in the market for commercial loans immediately after the disaster and during the lingering aftermath. Our interviews suggest that, most often, local jurisdictions deal with banks or credit unions that have a serious stake in the community in that the financial institutions either are locally owned or have a long history of community involvement. It is typically to the advantage of such institutions to have the community and the local economy recover quickly. We saw at least one instance in which a local bank worked with the local government for more than a decade until that government regained its financial footing.

The federal government is also a source for operating loans. The Department of Homeland Security, through FEMA, will provide a community with a loan through its Community Disaster Loan Program.[24] As of the time that this is book is going to press, the loan is not to exceed 25 percent of the government's operating budget for the fiscal year during which the event has occurred, and then only to a maximum of $5 million. It is possible that the maximum loan can be increased beyond $5 million, and under extraordinary circumstances, these loans may be forgiven. The big problem for local government is that the financial emergency almost always lasts for more than one fiscal year, and the needs almost always exceed 25 percent of the local government's operating budget and are almost always in excess of $5 million. It would be prudent, then, for local officials not to count on federal loans for much help in getting through the financial crisis.

Institutional Arrangements within States

Florida pioneered an approach to help local governments with their revenue needs following disasters. After Hurricane Andrew, the state helped jurisdictions in Dade County recover some of the sales taxes they would have received had it not been for the hurricane. Rebuilding efforts in South Dade

County generated significant sales outside the damaged area, where sales tax revenues rose in excess of what they would have been had the hurricane not occurred. For three years, the state sent those excess sales tax revenues to stricken local governments in Dade County. It was a good start except that, from the standpoint of Homestead, Florida City, and other South Dade jurisdictions, three years was not enough time for them to recover to the point where they were able to generate sufficient revenue to operate at an acceptable level.

Nonetheless, it seems entirely reasonable for local jurisdictions to prevail on governors and legislators to create programs in which unexpected revenue gains from disasters are used to provide assistance to the disaster-hit localities. Similarly, if neighboring communities are burdened by persons displaced by the event, they, too, should receive a share of the unexpected revenue gains to help offset their increased costs.

A second way that states can help municipalities is to enact "hold harmless" laws for other centrally collected, locally shared revenues, particularly when the distribution of those revenues is formula driven. In a few states, for example, school districts that suffer large and immediate reductions in school enrollment after a disaster are protected from rapid reductions in state aid for schools. The laws vary, but the general idea is that the state, in allocating aid, uses the previous year's enrollment figures, thereby providing assistance to a district that has been affected by the disaster. If the population loss is more or less permanent, annual reductions in assistance are phased in gradually until the district has balanced its resources with its enrollment. In Iowa, the state provides additional support to districts that are serving the displaced students.

FUNDING FOR LONG-TERM RECOVERY

Most of the money made available to a local government after a disaster seems to be given in the belief that community recovery will occur more or less automatically once provisions are made for temporary assistance to those who lost their homes and livelihoods, funds are earmarked for removing debris, and support is there for replacing public buildings and infrastructure. We believe those things to be necessary but not sufficient for long-term recovery. For that, as we see it, local officials are left mainly to their own devices to cobble together funds to help ensure long-term community viability.

Federal Contributions and Grants

FEMA is not responsible for financing a community's long-term recovery. Under the Stafford Act of 1988 (the amended Federal Disaster Relief Act of 1974), it is responsible for providing funds sufficient to rebuild local government facilities to where they were before the event. Although we have yet to see an instance in which the rebuilt facilities are not far superior to those that were destroyed, FEMA nevertheless has no mandate to facilitate the long-term recovery of the community system, including its economic base. Fortunately, the Department of Housing and Urban Development (HUD) and the Economic Development Administration (EDA) of the Commerce Department do have responsibilities for long-term community recovery, and both provide significant amounts of funds for approved projects intended to stimulate local recovery.

The municipalities that fared best in receiving financial grants from federal agencies like FEMA, HUD, and EDA shared several important characteristics. First, almost all of them hired, contracted with, or designated a very bright, aggressive person or persons to learn all that was needed to know about available federal programs and eligibility requirements. Second, they ensured that they were well versed in program requirements and regulations. Third, their local officials were adept at quickly drawing up sensible, readable project proposals to the various granting agencies. Fourth, each successful jurisdiction had staff designated to do little other than work on grant applications and administration, a job that was considered to be just as critical as bringing the water purification and distribution system back on line. Fifth, the jurisdictions did not blame the federal government or individual federal officials for their problems and for seemingly unnecessary red tape; instead, they worked with individual federal officials to provide the required information quickly and accurately. If they found a particular official hard to work with or unreasonably demanding, they did what was required anyway, at least until an opportunity arose to have that official replaced. Finally, successful local governments had sophisticated accounting systems in place that enabled them to track every dollar spent—to show work orders, bid sheets, time cards, and all other documentation required by the federal government to ensure reimbursement. Moreover, their records were sufficient to accommodate the needs of federal auditors who invariably arrived years later: municipalities told us of project files that were not audited until up to seven years after the project and of files not closed until almost ten years later.

Local governments without high-quality accounting systems and staff found themselves in trouble at almost every turn. Some cities were forced to return large amounts of grant money given to them because they failed to comply with the regulations. The time required to fend off the auditors was, we estimate, much more than the time required to do it right the first time.

Technical Assistance

Technical assistance is important to augment local talent and specific knowledge about methods and sources of additional financial assistance. Local governments are individually responsible for developing the strategies and action plans that will help their communities work toward recovery, but several states provide exemplary technical assistance to local communities in that effort. Florida, which did very little to assist local government in the wake of Hurricane Andrew, has since become the national leader in providing technical and financial assistance to local governments in communities that experience disasters. Minnesota and Georgia also provide superior assistance to communities that ask for help in developing and implementing recovery strategies.

The Los Alamos Anomaly

The Cerro Grande fire in 2000 was started by the U.S. Forest Service in the Bandolier National Monument. The intent was to conduct a planned burn to reduce the amount of fuel on the forest floor, thus reducing the risk of massive fires such as those that had ravaged New Mexico in prior years. However, the fire got away from the Forest Service and burned nearly 48,000 acres. According to the Governmental Accountability Office, it burned forty structures at the Los Alamos National Laboratory and 235 residential buildings, displacing 400 families.[25] Many of the residential buildings were multifamily, some left over from when the Army built them in the 1940s for the Manhattan Project. The residential buildings averaged fewer than 1.5 households per building.

Shortly after the fire, Congress enacted the Cerro Grande Fire Assistance Act, which provided an unprecedented amount of aid to the community of 18,000. FEMA provided "nearly $570 million in disaster expenses and claims paid to individuals, businesses, communities and tribes."[26] Assistance amounted to more than $31,000 for every man, woman, and child in Los Alamos County. It totaled about $1.4 million for every household displaced and something over $2 million for every residential structure burned. For perhaps the first time in disaster assistance history, businesses were compensated by the federal government for business they might have had had the disaster not occurred. Local business persons were compensated for the period they were closed based on their earnings during the same period in the prior year.[27]

It is difficult to imagine that none of the homeowners or renters had fire insurance, that none of the business owners had business interruption insurance, and that other agencies did not provide funds to the residents of Los Alamos. No one died from the fire, and while everyone can certainly empathize with those who lost their homes and contents, the amount of money spent to compensate the residents seems excessive when compared with assistance provided almost anywhere else. In the singular case of Los Alamos, special legislation granting large reimbursements to individuals and businesses appears to have been possible because of the acknowledged culpability of the U.S. Forest Service in igniting the blaze, the federal government's desire to avoid lawsuits, and a well-placed congressional delegation. Importantly, we have not found another case like Los Alamos, and we discourage any jurisdiction from thinking that its disaster will yield similar reimbursement results.

REBUILDING THE TAX BASE

Of course, long-term recovery of both the community and the local government serving the community requires either rebuilding or developing anew a base of wealth, income, and business activity that is adequate to provide the needed revenue for collective action. Unless that is accomplished, when the federal aid ends, the community and the local government will wither.

In some communities, the tax base rebuilds itself in short order. In others, where the economy had been sluggish before the disaster or became unstable following it, establishing a viable local economy in the aftermath of the disaster almost always becomes the primary objective of local officials and community leaders. In our experience, most attempts to build a dynamic local economy and an adequate tax base in those communities relied on the traditional economic development approaches: creating industrial parks, advertising for clean industry, providing all the tax breaks permitted by law, and endorsing chamber of commerce efforts. We saw little evidence of local government officials and their business and industrial colleagues working to develop a development strategy based on a rigorous analysis of what might be possible, of what basic investments in infrastructure or education might be required, or of what regional strategies might be employed to help improve the area economy in general.

Establishing, reestablishing, or enhancing the local economy is critically important regardless of whether a community has suffered a disaster. We continue to look for success stories that embody lessons for other communities, and when we find them, we will do our best to add those lessons to our understanding of effective management for long-term community recovery.

PART 4

WHAT WE LEARNED ABOUT LONG-TERM RECOVERY

FROM OUR DISCUSSIONS WITH CITY managers, top staff members, and other officials in more than a score of disaster communities, we learned that each community has a unique context and that, for each, recovery will mean something slightly different, require somewhat different choices, and take a somewhat different path. We also learned that some considerations and concerns are common to almost all communities that experience a disaster. We have compiled what we learned into the following chapters to address what seems to help with recovery and what does not. The sequencing of these chapters should not be taken to imply a strict sequencing of actions. It is likely that many of the activities will be undertaken at the same time.

CHAPTER 13

RESPONDING TO THE EMERGENCY

"One of the true tests of leadership is the ability to recognize a problem before it becomes an emergency."

Arnold H. Glasgow

Street cleanup in progress. Photograph taken in New Orleans in October 2005.

*B*EFORE AN EXTREME EVENT, LOCAL emergency management centers are activated, and the local government goes on high alert. After that, and both during and immediately after the event, local government activities are usually driven and prioritized by what officials see as necessary: warning residents of impending peril, coordinating communication, controlling crowds, helping with evacuation, monitoring flood levels and sandbagging as flood waters rise, dispatching rescue teams, ascertaining damage, shutting down public utilities as needed, detecting and suppressing fires, establishing crime scene perimeters, and working to ensure public safety.

Of all the things that local government does in connection with a disaster, immediate response is perhaps what it does best. We have little doubt that this is because local government agencies train extensively and have considerable experience with response. Essentially, all the local governments we studied engaged in a common set of response activities in the immediate aftermath of the extreme event.

RESTORING GOVERNMENT SERVICES AND REPAIRING PUBLIC FACILITIES AND INFRASTRUCTURE

Within the overall set of activities that must be undertaken before, during, and after a disaster strikes are those that are usually initiated almost immediately. These include establishing basic public safety and rule of law as well as removing debris generated by the extreme event, initiating repairs on government owned infrastructure and restoring public utilities services for which the local government was responsible before the event. No one wants to invest in a community where either they or their investment is in imminent danger of fire, flooding, massive traffic problems, violent and lawless gangs, or extensive criminal activity.

These activities signal the beginning of the recovery effort. Local government crews repair streets and bridges; restore water purification and distribution systems; put sewage collection and treatment facilities back into operation; and fix utility generation, distribution, and collection facilities (if these facilities are owned by the local government) or let contracts to private firms to do so. However, local governments typically do little or nothing to rebuild infrastructure or facilities they do not own. If the county government owns a hospital or other facilities, it is up to the county to get that work done; similarly, state governments are responsible for repairing or replacing their damaged facilities.

In some communities, which governmental entity—city, county, or state—removes the debris deposited by the flood or tornado from various streets in a community depends on whether the street is designated local, county, state, or national. Sometimes, but not always, those arrangements are made before an extreme event. Of course, it is best not to wait until streets are clogged with debris to work out who is to do what: officials in communities that had not made out arrangements in advance told us that working out the details of who is responsible for what *after* the event has occurred is time-consuming and can generate friction.

Deciding which elements of the built environment to tackle in what order and for which parts of the local community is no easy task. While it may seem obvious that the first elements requiring attention are those associated with power and potable water, it is not always possible to address those elements first. Streets may need to be at least partially cleared and buildings may need to be quickly inspected before power can be restored. Officials may find themselves in the unenviable position of having to decide which parts of the community will receive services first according to which parts they anticipate will most likely be repopulated first.

PROVIDING SHORT-TERM ASSISTANCE TO INDIVIDUALS AND FAMILIES

In concert with other public and private organizations, local governments provide short-term assistance to residents. They help recover the dead, and provide morgues or temporary storage for the bodies so they can be identified. For the survivors, they help to provide medical treatment, temporary shelter, food, and other resources. To do all this, they receive help from nearby local governments, private utilities, state agencies, federal agencies, nongovernmental organizations, and private individuals.

But sometimes local governments are unable to provide needed assistance. In those cases, they often get assistance from nonprofit organizations and, sometimes, from individual citizens. We learned that in some of the communities we visited, private citizens provided hot meals to those who suffered losses or were working on the cleanup. "The power had been out for some time and the food in our freezers had begun to thaw," one small-town restaurant owner said. "I would much rather cook it and give to those who need it than throw it away." In many communities, local volunteers and volunteers from nearby and distant communities arrived simply to help clean up and get people back in their homes.

Some county governments and a few communities provided counselors to help with emotional and psychological problems. Some county governments offered counseling during the first few months after a disaster; however, emotional problems tend to emerge later rather than earlier. Federal funds are now available for local governments and state agencies to provide longer-term counseling, but the funds must be requested early in the recovery process, perhaps before the need has become apparent.

HELPING RESIDENTS RESUME ROUTINE ACTIVITIES

In the aftermath, virtually everyone and every organization will be working hard to return to pre-event normality. Residents want to return home. Business owners typically want to reopen as soon as possible, regardless of whether there is any immediate market for their goods or services. Some activities are more urgent than others and deserve a higher priority in local government recovery efforts—namely, the needs of disaster workers and remaining residents. This means ensuring the availability and accessibility of food, fuel, housing, pharmacies, financial institutions, and building supplies.

Essential Goods and Services Must Be Available

Priorities have to be set to help ensure that the most urgently needed goods services can be obtained as needed. Some people can camp out for a while in the ruins of a community, but for sufficient numbers to return or to stay, essential goods and services must be available. Local government cannot be responsible for all these goods and services, but it can facilitate their availability.

There must be fairly reliable power, water supply, and sewer service for people to stay or return, for hospitals and health care clinics to operate, and for businesses and industry to reopen. Pharmacies must be open for those who need access to medications and medical care products. Food has to be available on a regular basis, which means that at least some restaurants and grocery stores have to be in operation. For those that will require inspections beforehand, the local government must be ready to expedite inspections and permits. Fuel is needed, so gasoline stations have to be open and accessible to suppliers and customers; main transportation routes have to be open and available as well, and some public transport may be necessary.

General-purpose local government has to be operating, at least at minimal levels, for routine governance as well as to address recovery problems. At least some schools must be open and operating.

Housing Must Be Available

At the same time that basic services are being put back into operation, assistance is needed to move people from emergency shelters to temporary shelters and finally to permanent housing. FEMA trailer parks and recreation vehicles in front of some owners' homes will meet some of the need for temporary shelter. Low-cost hotel and motel rooms must be available for those who are in the community temporarily for cleanup and construction. So that enough workers are on hand to help jumpstart the local economy, the community must move promptly to ensure that there is sufficient affordable housing available that is relatively convenient to workplaces. If housing is available in nearby communities, businesses can help local government organize bus transport and carpools for those who have jobs in the recovering community but do not yet have housing in it. Such housing need not be permanent; permanent housing will eventually be provided to meet the effective demand.

Some larger employers may be able to work with employees and prospective employees to create living arrangements that are suitable, at least, for the short term. In New Orleans, for example, where housing accommodations were difficult to secure, most of the hospitals that reopened provided temporary housing for their employees for months after Katrina. Those hospitals that were closed continued to pay employees for several months even though employees could not report to work; doing so enabled employees to pay rent or mortgages. And several employers in the city paid for employees to commute daily from Baton Rouge.

Before long, organized efforts will displace spontaneous efforts by individuals and small groups. In New Orleans, FEMA brought in trailers to provide temporary housing that was a little more permanent, and housing assistance continued for an extended period. But the length of time that individuals need "temporary" assistance varies considerably, and few local governments do much to help individuals and families over the longer term.

Usually, the extent to which local, state, and federal governments help private individuals and organizations repair, rebuild, or replace facilities is a function of the kind of enterprise it was and the form of its organization. Loans and small grants are generally available for homeowners to repair their homes. Some local governments buy land, create subdivisions, and sell houses on the private market; others obtain vacant lots and housing that have been foreclosed for nonpayment of taxes, rehabilitate the housing, and sell it to people who have lost their homes in the disaster.

Beyond assisting homeowners, however, the government's role in helping private parties restore productive parts of the community is limited to a few kinds of organizations. For example, the federal government will provide funds to repair damage from extreme events to hospitals owned by state and local governments. Sometimes it will provide money to those hospitals for hazard mitigation. It may do the same for privately owned hospitals if they are organized as nonprofits. But it will not provide financial assistance to privately owned, profit-seeking hospitals that provide exactly the same services and that are also required by law to serve uninsured and indigent persons.

Similarly, privately owned, profit-seeking local businesses with disaster-related losses are eligible for very little, if any, assistance. Many self-employed small-business owners told us that they were refused assistance from local and national philanthropic organizations even though they, too, were resident homeowners with substantial losses. Had they been employed by someone else, they would have been eligible, but being self-employed precluded them from the benefits that others received.

On the other hand, if the business is agricultural, more than a score of federal programs provide grants and loans to private, for-profit owners to cover direct losses from a wide variety of extreme events. Presumably, the congressional rationale is that a privately owned, profit-seeking farming operation is more worthy of assistance than a privately owned, profit-seeking firm that focuses on manufacturing or distribution, or even one that provides health care.

Money Must Be Available

After the immediate emergency has passed, recovery can begin when up-front money is available to the local government from both within and outside the community to undertake essential tasks. The money might come from reserves, insurance, loans, or grants, but without it, almost nothing can be accomplished. Private organizations—both for profit and nonprofit—need money, too, if they are to rebuild, repair, replace, pay employees, buy supplies, and so forth. Last but not least, individuals and families need money to meet immediate needs and to initiate their own recovery activities. Individual recovery efforts, of course, when taken in aggregate, are critical to community recovery.

Having money available means that local financial institutions must be operating, at least at a makeshift level. People and firms need access to their accounts, and local financial institutions in the communities we studied worked hard to open quickly. We have seen them operating in warehouses, trailers, and make-do facilities when their own buildings were unusable. And we have seen bankers in hip boots shoveling cash into biohazard bags for subsequent shipment, counting, and decontamination or replacement.

CHAPTER 14

GOING BEYOND EMERGENCY RESPONSE

"You may be disappointed if you fail, but you are doomed if you don't try."

Beverly Sills

Many communities find it important to remember and commemorate the efforts of residents to rebuild the community and to achieve a sense of normalcy, even though they may have redefined what is normal. Photograph taken in Grand Forks, North Dakota in May 2004.

W HY, WE WONDERED, DO SOME local governments become more involved in community recovery than others, going far beyond the typical set of efforts described in the previous chapter? We can't provide a definitive explanation, but we can provide some clues that emerged from our studies.

PERCEPTIONS OF THE CONSEQUENCES OF THE EXTREME EVENT

Disasters can be viewed as "socially constructed" events. Objective measures of loss and consequences do not make a disaster a disaster, but the collective interpretations of its consequences do.[29] That is why an airplane crash resulting in 300 lives lost is typically viewed as more tragic than the automobile accidents that take 50,000 lives annually. Proximity, similarity, and national attachment are also important in perception. For some Americans, the loss of roughly 2,000 lives in Hurricane Katrina, for example, was more devastating than the loss of 125 times as many lives in the 2004 Indian Ocean tsunami or the deaths of nearly 100,000 persons each in the 2008 typhoon in Myanmar and the Sichuan earthquake in China.

It appears to us that local governments take on a more activist role in disaster recovery when they believe that the event affects a very large proportion of the community very deeply, and when it is evident to local leaders that economic or social recovery is unlikely without aggressive and concerted action to facilitate or stimulate it. In other words, leaders are more likely to intervene when the community does not behave the way a self-organizing system is expected to behave without that external intervention. Often, the purpose of such intervention is to jump-start the community's return to being a viable self-organizing system.

The city of Los Angeles and Los Angeles County certainly did a lot following the 1994 Northridge earthquake, but their efforts were generally limited to restoring services, repairing infrastructure, and assisting with repair and replacement of lost housing stock. Despite the enormity of loss, the disaster did not threaten the solvency of local governments, nor did it pose severe threats to the continued viability of communities in the Los Angeles metropolitan area. However, in smaller cities that are not embedded in a megalopolis where losses are not large by regional or national standards, the consequences of an extreme event are perceived as affecting almost everyone in the community and posing threats to continued community viability. In such cases, recovery becomes the issue on everyone's "front burner." It dominates talk in the local breakfast haunts, the front page of the local paper, every nightly newscast, and every politician's agenda. Although they probably do not think about it in this way, it becomes a priority for individuals to try to figure out how they can individually and collectively regain their pre-event status.

POLITICAL CULTURE, POLITICAL REALITIES, AND POLITICAL PRACTICALITY

"Political culture" is a term used to describe how most people in a particular place tend to perceive the appropriate role of government and collective action. Political culture varies from place to place: within regions and states, and even within various parts of individual states, dominant ideological beliefs and pragmatic concerns about the appropriate role of government affect who gets elected to public office and what gets done. Such beliefs and concerns also influence the actions that state and local governments take in the realm of disaster response—how proactive they are toward hazard mitigation, how involved they are in post-event recovery, and how open they are to a wide range of recovery strategies. In some communities, the local government uses government resources to build subdivisions, acquire and sell lots for development, and even build and sell housing directly to end users. In other places, that role is left entirely to private business. Some states and communities enact stringent building regulations to help ensure the safety of residents; other states do not even require every general-purpose local government to have a building code. Similarly, government efforts to ensure compliance with codes vary.

Sometimes political culture mixes with political reality. Florida City, Florida, is a small, predominantly African-American community adjoining Homestead. Florida City aggressively tried to attract business in the post-Andrew months and years. It acquired building lots that Dade County had taken because of tax delinquency, improved those lots, and sold them to local families who qualified for them. The small city was proactive for a number of reasons; most noteworthy was the fact that the mayor had been in office for many years and enjoyed the support of the city council, and this political continuity positioned the mayor and council to take bold actions. Many other communities do not have long-term incumbents with relatively little competition for elective office; in addition, they may be more politically divided or divisive. In those communities, active issues that divide constituents and officeholders may reduce the likelihood of aggressive, sustained developmental initiatives.

Political practicality can affect the amount of money that is available to local governments for recovery. Princeville, North Carolina, founded by freed slaves in the 1880s,[30] is a small community with an almost entirely African-American population. After Princeville was badly damaged by Hurricane Hugo in 1989, President Bill Clinton visited the town and promised that the federal government would rebuild it; presumably, he wanted to make a statement about the government's commitment to poor and destitute members of a racial minority. It also helped, however, that Princeville was small, so it could probably be rebuilt for a lot less money than it would take to rebuild New Orleans or even Homestead. Thus, substantial funds flowed into Princeville for a variety of projects, including public housing.

As it happened, however, the struggle to make Princeville a viable community was a daunting challenge even before the town was ravaged. Princeville's starting position was behind the "start line." Its economy had been in tough shape before the extreme event, especially since nearby communities were thriving and drawing away the town's residents and money. Existing drug and crime problems escalated with the introduction of public housing. New people who moved into the community had fewer skills. Projects were started and left unfinished. Thus, the efforts to rebuild Princeville and turn it into a thriving community after it was badly damaged by Hugo is a case study in how good intentions and lots of money—in this case, massive federal assistance—sometimes aren't enough to secure a community's recovery.

THE RESOURCES AVAILABLE

It is impossible to do much about recovery without financial resources. Thinking of the remote, impoverished places in Southeast Asia that have been destroyed by typhoons and earthquakes just in the last decade, it's hard to imagine how such communities can ever recover. Communities in the United States are fortunate that the federal government provides so much money for disaster relief and recovery, that private individuals are extraordinary generous, that so many strong and generous nongovernmental nonprofit organizations exist, and that, overall, the economy generates sufficient excess to provide assistance.

Sometimes, however, it seems that what local governments do in the attempt to foster community recovery is driven solely by the availability of federal and, to some extent, state monies for specific projects. Block grant funds from the Department of Housing and Urban Development, for example, provide a source of revenue for projects designed to aid the specific community. Because they are block grants, the funds may be used for any of a wide range of programs and activities. And because local governments are almost always extremely short of funds, and because their revenue from local sources usually shrinks in the aftermath of a disaster, block grants seem to induce local officials to do things that they would not otherwise do: for example, creative officials may use the funds to create synergies with otherwise apparently unrelated projects that are otherwise funded through narrowly specified sources of assistance, or they may use the block grant money to pay for an activity on which they might otherwise have spent local government funds so that they can put their own funds to another use. It should be noted, though, that state and, especially, federal programs help fund activities that almost all local governments would have to undertake regardless of whether federal funds were available—including, for example, debris removal and repair or rebuilding of damaged public infrastructure. However, if a granting agency wants to discourage a local government from substituting the grant for local money that would otherwise have been spent on that same activity, a matching grant is the preferred instrument. Under most circumstances, matching grants induce the recipient government to make additional expenditures.

Sometimes restrictive conditions are placed on the use of funds, and this done for apparently two reasons. The first is that federal policy makers want to encourage a local government that has been hit by a disaster to undertake those particular activities that they believe must be done to assist in the recovery effort. The second, it seems to us, is that officials at any one level of government distrust what officials at another level might do with monies given to them *without* having strings attached. So instead of providing unrestricted funds, the granting government gives funds for specific projects and activities without giving much thought to how those various projects might contribute to an overall local strategy. Local officials are generally left to cobble together a sensible overall plan from a generally unrelated host of projects for which assistance is available. In a few communities, however, local leaders created recovery plans and then searched for federal and state grants that would provide financing for various components of their strategy.

If local governments are not likely to spend money for some activities after a disaster unless they get federal assistance, they are even less likely to spend money on mitigation activities—activities intended to reduce the vulnerability of the community to subsequent events—*before* a disaster unless state and federal funds earmarked specifically for that purpose are available. Even with incentives to mitigate, however, not all local governments avail themselves of the opportunity to reduce their vulnerability even though they would be reimbursed for doing so. Some are in denial about possible hazards. Paradoxically, in some communities an extreme event will cause people to believe that they are now virtually immune to another extreme event, at least for the foreseeable future—almost as if experiencing an extreme event "inoculates" a community against similar events. Such beliefs are reinforced by people's failure to understand risk. For example, after experiencing what is described as a "100-year flood," some people will erroneously believe that another such flood won't occur for another 100 years. When asked to explain why he was rebuilding his home in Chalmette, Louisiana, which was completely inundated by flooding after Katrina, one resident said, "I just don't think it's going to happen again—something like Katrina happens only once in a hundred years. By that time, I'll be dead."[31]

Lack of awareness about mitigation opportunities also interferes with mitigation activities. People may fatalistically believe that no amount of mitigation will protect their communities. Or they may believe that the cost of mitigation outstrips the benefits. Or they may not perceive that they have the resources (financial, human, capital) needed to implement mitigation activities.

Finally, mitigation activities compete with other activities on the local government agenda. Unless funding or some other inducement draws their attention to mitigation, we should not be surprised that local governments will attend to whatever is most urgent and obvious—characteristics that mitigation tends to lose with the passage of time after an extreme event.

THE CREATIVITY AND RESOURCEFULNESS OF
LOCAL LEADERS

Some local officials and community leaders are simply more creative and resourceful than others when it comes to cobbling together resources from sources beyond what the local tax base generates. Vision, creativity, and persistence are powerful forces for garnering funds from diverse sources to help a community rebuild and revitalize. Local governments that went out of their way in the aftermath of a disaster to hire people with special skills in putting together plans and proposals were usually able to do more and receive more governmental assistance than those that did not.

CHAPTER 15

LOCAL GOVERNMENT OPERATING SYSTEMS MUST BE OPERATIONAL

"If I had eight hours to chop down a tree, I'd spend six sharpening my axe."

Abraham Lincoln

The wholly unusable city hall in Gulfport, Mississippi, following Katrina's storm surge. Photograph taken in May, 2006.

*I*N THE IMMEDIATE WAKE OF a disaster, local government administration is anything but routine. As the emergency begins to wind down, it is common for administrators to breathe a sigh of relief. Too early! The aftermath of the disaster will be even more complicated and trying than the emergency response. This chapter identifies actions that local government can and should take immediately after the disaster to prepare itself for recovery efforts.

This is not a step by step checklist for getting local government apparatus back in good working order or prepared to take on the massive task of recovery. Instead, it is a collection of things that we learned from officials who have been through one or more disasters.

FORTIFY ACCOUNTING AND FINANCE SYSTEMS

Almost every local official told us that accounting and finance systems are put to the test in the aftermath of a disaster. Federal project funds flow into jurisdictions when the president declares a disaster there. But after the money is gone and the projects are completed, the federal government's program auditors arrive—looking for documentation of eligible expenditures, for expenditures that are ineligible under the terms of the grants, for errors and mistakes made by local officials, and for assurance that procedures were followed precisely. These auditors want (demand) not just complete records on how every dime was spent but compelling evidence that every dime was spent in compliance with program regulations. If they don't get that evidence, they will want to get some or all of the money back. Some local governments find themselves facing the prospect of having to write a large check to the federal agency.

It is not simple to keep track of expenditures after an extreme event. The local government staff is dealing with situations that it has not previously encountered, trying to do the job that it did before the disaster while dealing with new demands that it has never faced before. Moreover, most federal program rules are complex, and not all agencies employ the same procedures. In addition, federal agencies may hire temporary personnel in response to a specific disaster. These personnel are often less knowledgeable about rules and regulations than longtime employees, which may explain why local officials in several cities reported having received contradictory information from agency personnel on successive days.

Experiences varied. Some local governments had wonderful dealings with federal officials; others did not. Some small municipalities hired one or more individuals to read and understand the regulations of every federal program with which the local government was or might be involved; those localities had fewer difficulties over the long run. Grand Forks, North Dakota, for example, had help from consultants when setting up their systems and thus fully met federal expectations. But local governments with barely adequate financial accounting systems and documentation procedures before the event often found themselves in serious trouble when state and federal agencies conducted their reviews. Local governments with insufficient systems have spent years fighting federal efforts to recapture tens of thousands and sometimes millions of dollars. Even those with adequate systems can find themselves answering questions from federal auditors five, six, or seven years after the disaster. The lessons are clear: from the very first day, local governments must make sure that *complete record-keeping and adequate accounting systems are in place.*

Local governments benefit from having a chief financial officer on staff. The federal government doesn't advance funds; it reimburses after expenditures have been made. In some states, the federal money goes through a state government agency before it gets to the local government. Delays in reimbursement are common. Managing cash flow and arranging for bridge financing can become a staggering problem. Maintaining an overall view of local government finances and keeping the jurisdiction within safe financial limits requires more than accounting skills; it requires someone who understands finance, management, intergovernmental relations, and, in general, how things work in a political system.

REDUCE NONESSENTIAL BUREAUCRACY

In government and other monopolies, there is sometimes a tendency to externalize costs to the user. For example, in most states residents see the Department of Motor Vehicles as externalizing its administrative costs to the customer, who gets to wait in line so that the agency doesn't have to spend more money on employees. Local government is sometimes in a monopoly position: where else can someone get a building permit? But unnecessary steps in processes usually create more hassles and, consequently, higher costs, and in the aftermath of a disaster, residents should not have to bear the burden of such unnecessary hassles and costs. Nor can a local government afford to generate the additional frustration and animosity among residents who need help to speed their personal recovery. Angry citizens will contact their city council members, who can make life extraordinarily difficult for local government staff members.

Florida City, Florida, a small, poor jurisdiction immediately adjacent to Homestead, has struggled for a long time to build a viable local economy. As described previously, it developed a particularly useful approach after Hurricane Andrew. The mayor and his chief associates decided to slash red tape when dealing with firms that were thinking about locating within the municipal boundaries along U.S. 1 at its junction with the Florida Turnpike. Whereas some municipalities require prospective firms to meet with three or more committees and to comply with the conditions imposed by each, Florida City officials decided to simplify the process: the mayor and two department heads met with the prospective employer at a time and place convenient to the prospect, and the three were empowered by the city council to cut a deal with the prospect essentially on the spot. Eliminating the red tape worked well for Florida City—prospective builders and employers liked the service and the speed of decision making—but such an approach requires trust that the small team will work toward the interests of the community as a whole and not use the relaxed process to line their own pockets. Florida City's fairly homogenous demographics and political stability facilitated this approach.

Local governments need to recognize that it will be very difficult to conduct business as usual in the wake of a disaster. While the usual expectation is that they will create and adhere to standardized operating procedures that ensure stable and consistent outcomes, their job in the aftermath of an extreme event is to get needed work done as quickly as possible and solve problems, not to offer a complex array of services that generate paperwork. They can return to managing in ways that promote consistency when the external environment has stabilized. In the meantime, though, they need to be responsive and flexible. Because the overall job changes after a disaster, daily operations must change accordingly.

From our research, we have concluded that recovery requires a local government to be flexible and adaptive; it must be able to do things it has not done before, and do things in ways it has never done them before. And it is far better to simplify processes beforehand, when there is time to give them serious thought, than it is to waive them or ignore them in time of crisis. In the aftermath, there are not enough resources to do things that don't need to be done or that can be done more simply, and there is certainly not much time for contemplation.

ENSURE ADEQUATE STAFFING

Municipalities experienced staffing problems as they began to work toward community recovery. First, staff members tended to become exhausted from the long days, from sleeping at city hall, from worrying about how they could help with recovery at home while being at work all the time, and from the 24/7 regimen that many lived for months on end. They needed backup.

Second, many local governments found that they did not have enough of the requisite skills—or they had inadequate numbers of people with those skills—to deal with the huge and complicated task of rebuilding the community's infrastructure, writing project proposals to state and federal agencies, and managing complex projects. This was often the case with small and mid-size communities, which almost always had to augment their staffs after an extreme event by getting additional help in specific areas. Some local governments hired high-level temporary staff to help search for funding opportunities, develop plans and proposals, set up accounting procedures, understand legal implications, work with state and federal officials, and provide counsel to harried elected officials and managers. Where we found a local government that had done this, we found no one in that government who had regretted the choice. In some instances, the individuals who were hired for the short term were still there years later, still helping with complicated recovery problems.

We believe that many local governments typically have to augment their capacity in order to evaluate and assess systemic community consequences, to devise and evaluate recovery strategies and programs, to manage programs that involve and cut across several agencies and departments, to work with granting agencies to maximize the help available to the community while minimizing the hassles, and to track expenditures. Contracting for temporary help is often essential for dealing with the combination of heavy workload and new challenges in the aftermath of a disaster. Sometimes help is needed for contracting with dozens of vendors to remove debris, repair facilities, and design and build new facilities. That work is demanding and time-consuming, so it is often necessary to add trained staff to building, finance, and public works—agencies that face enormous increases in workload because of the disaster

Local governments that searched and recruited nationally for top-notch consultants usually fared better than those that tried to make do with what they had. Good consultants are those who have been through disasters before, have a solid track record of being helpful in other communities, are easy to work with, and will be there when they're needed. It is best to check with other local governments to learn about their experience with specific consultants or consulting firms. If you can see a problem coming, as in the case of major flooding on a long river, it makes sense to have made these inquiries before the flood waters reach your community.

It is important to remember that the federal government will cover the cost of people who are working on recovery, but it will not pay for people doing the routine business of running a city government. At least one local government we studied made the mistake of hiring everyone it needed for the extra workload as regular city employees. When the time came that the federal government would no longer pay for recovery workers, the city had considerable difficulty removing those employees from the payroll. Local revenue was inadequate to cover the operating budget with those employees on the payroll, and trouble ensued.

LEARN ABOUT ASSISTANCE PROGRAMS

As previously noted, some local governments in our study hired one or more individuals as consultants and charged them with responsibility for reading and understanding the regulations of every federal program with which the local government was or might be eligible. Where that was done, the local government was usually successful in obtaining grants and assistance with less difficulty than communities that chose to rely on an already overburdened staff.

A host of assistance programs are available from federal, state, and private sources. Some of these are available to local governments; others are available to private and nonprofit organizations. Rather than hoping that those nongovernmental organizations in the community find out which assistance programs are out there, local officials can work to ensure that every person and every organization in the community understands which programs exist, what their eligibility requirements are, what their basic rules and regulations are, and how to maximize the probability of getting assistance from them. A clearinghouse function for such assistance programs might be established in city hall, on the city's Web site, or in cooperation with a nonprofit organization within the community.

Disaster Assistance: A Guide to Recovery Programs (FEMA-229) is a guide to federal disaster recovery assistance programs. It describes thirty-four disaster-specific recovery programs and lists another fifty-nine recovery programs considered to be "disaster-applicable."[32] Some of these programs provide money to mitigate future losses in the community from extreme events; others provide funds for repair or replacement. Some provide funds to local governments; others provide loans and grants to individuals and to businesses. The local government that does not know which programs exist and what their eligibility rules are is very likely missing out on funds that it can use for recovery or to reduce losses from the next extreme event.

CREATE A HIGH-LEVEL PROBLEM-SOLVING TEAM

Since childhood, most of us have been subjected to motion picture story lines in which a demented scientist or some mindless monster or force of nature threatens a disaster that will—unless thwarted—end life as we know it. A single hero or heroine, usually a reporter, faces the threat and saves the world while the rest of us, especially scientists and government employees, stand helplessly on the sidelines looking for all the world like deer in the headlights. In real life, however, "superheroes" do not fly in and save the day. Instead, teams comprising public employees, private organizations, and citizens with exceptional skills, dedication, and perseverance are needed to address extreme events and their consequences. Real men and women, working collaboratively and over time, tackle the multifaceted issues that are associated with community recovery from extreme events.

The local government also needs a high-level problem-solving team, a team that cuts across specialties and across departments to see the big picture and to understand how the parts fit together. This has to be a team of people who are willing to tell top leaders when something doesn't appear to make sense, who have good ideas, who can work together, and who can handle responsibility. The team may be made up of department heads, but it doesn't have to be. It might be an ad hoc group comprising the mayor, city manager, and two or three others who are particularly competent.

The individuals who make up these leadership advisory teams have to address some very difficult questions and tackle some very complex tasks. Perhaps the first of these tasks is to assess damage. Assessing physical damage from an extreme event is relatively easy compared with identifying the consequences to which such damage will lead. Which components of the community were damaged? How badly? What other effects have spread or are likely to spread through the community? If the community suffered only losses to its built environment, then the challenge is to repair or replace those elements as quickly as possible. However, if adverse effects have already begun, or if the community is faced with the prospect of addressing pre-event problems as well as problems generated by the extreme event, the situation is far more complex and demanding. The challenge of fixing long-standing problems exacerbated by the extreme event is extremely tough and, sad to say, sometimes virtually intractable.

CHAPTER 16

IDENTIFYING WHAT HAS TO BE DONE NEXT

"The trouble with the future is that it usually arrives
before we're ready for it."

Arnold H. Glasgow

More than three years after Hurricane Katrina, Methodist Hospital in
New Orleans East remains shuttered. Its reopening has been linked to
a population increase in the immediate area, something that has not
yet occurred. Photograph taken in New Orleans in March 2007.

*W*HEN THE COMMUNITY EXPERIENCES SOCIAL, economic, and
political consequences in addition to damage to buildings and infrastructure, local governments usually find themselves called upon to
orchestrate long-term recovery activities. Some communities will be able to try
to reestablish the kinds of dynamics that existed before the event; others will
find have to create a viable economy and social structure almost from scratch.

What each community needs in order to move toward recovery is largely contextual. It depends on what the community was like before the event, what parts of the built environment were damaged or destroyed by the event, and how extensive the damage is. It also depends on the nature and extent of the consequences that are cascading through the social, economic, and political parts of the community.

ASSESSING THE NATURE AND EXTENT OF THE CONSEQUENCES

Regardless of their circumstance, local officials will need to identify the problems facing the community and the steps that need to be taken beyond the initial response to the event.

Table 3. Losses in Eastern North Carolina from Flooding, 1999.

Numbers from Eastern North Carolina

1	School in Edgecombe County destroyed
3	Airports flooded
21	Wastewater treatment plants flooded
30	Downtown areas flooded
38	Drinking water systems damaged
40	Dams ruptured
47	Swine waste lagoons inundated with water
51	People killed in flooding
224	Coffins disinterred from cemeteries
250	Horses killed in the flooding
1,000	Roads washed out from flooding
2,000	Cattle killed in the flooding
15,000	Homes uninhabitable after flooding receded
28,000	Hogs killed in the flooding
30,000	Farms damaged
43,000	Homes damaged from flooding
69,717	People applied for emergency assistance
2,800,000	Turkeys killed in the flooding
4,200,000	Acres flooded in eastern North Carolina

Source: Tarboro (North Carolina) Daily Southerner, November 12, 1999, Flood–Special Edition, 9.

Identifying the More Obvious Consequences

After a disaster, it is important for a local government to obtain and interpret information about what has happened to the local community. It needs to understand the disaster at multiple levels and from multiple perspectives, including those of individuals, neighborhoods, consumers and businesses, infrastructure, and organizations. It needs to gather the massive amounts of disjointed information and work to interpret what it all means. What goods and services are not yet available? What is the housing situation, and where are people living? How have the large employers fared? What specific needs do they have? The local government needs to fully understand the information that makes sense and to integrate it with conditions that have yet to be completely understood.

We think it is critical to look for systemic consequences to the community as they begin to manifest themselves. Shortly after the event, local officials will likely have inventoried the immediate and more obvious consequences. One such inventory from a flood in North Carolina was quite specific; the list, as shown in Table 3, appeared in Tarboro's *Daily Southerner*.[33]

Moving Beyond the Immediate Consequences

Unlike injuries and deaths and damage to buildings and other structures, systemic community consequences will probably not be obvious in the immediate aftermath of a disaster. There is always a temptation to think that after the event occurs, the consequences are essentially immediate, and then you begin to address them. But consequences often continue to unfold over weeks and months, sometimes without much warning. In that sense, the disaster is dynamic. And since the consequences continue to unfold through time, the community must be monitored regularly to watch for them as they appear and to assess how they are playing out. Addressing these consequences is key to recovery.

Thus, local officials have to find ways to identify the consequences as they are unfolding. It requires continuous communication with people throughout the community, given that the view from city hall doesn't catch everything that deserves attention. Community leaders, business owners and managers, managers of financial institutions, executives in nonprofits, educators, citizens in the street, and officials in adjoining or overlapping units of government can all provide important intelligence on various cascading consequences and unanticipated events.

Local officials do not really need to spend time thinking about the ripple effects that are manifesting themselves in other communities: there is enough for them to focus on in their own communities. However, state and federal officials should pay close attention to those ripple effects. In New Orleans, a few of the hospitals that closed more or less permanently following the flood had been designated as the primary facilities to care for the indigent. Without them, the indigent sought health care at hospitals outside the community—hospitals not designated by the state to care for indigent people and for which there were no arrangements to reimburse them for those services. Similarly, cities that bear heavy burdens in the aftermath of disaster because they are inundated by people who flee the disaster site, as happened in Baton Rouge, should not be obliged to carry that burden by themselves.

Local officials should make an effort to identify any ripple reverberation consequences that begin to manifest themselves in the community. If other places are having to assume roles that the disaster-stricken community had formerly undertaken, it may indicate that steps need to be taken to cope with the apparent loss of key functions and roles in the community.

An intelligence function is also essential to identify and evaluate options and opportunities that may emerge in the community after an extreme event. Under normal circumstances, it may be sufficient to wait for these things to come to the attention of local government through the normal course of business, but unusual times demand unusual practices. In this difficult and critical time, it is more important than ever for local officials to keep their fingers on the pulse of the community.

SHAPING THE POST-EVENT COMMUNITY TRAJECTORY

Most local governments we studied worked to put things back the way they had been before the disaster. A few, however, took a different view. They believed it was important to develop a new or revised vision for the community and a basic recovery strategy that would help them realize their goal. Most were interested in developing a more vibrant economic base than they had before, with good prospects for the future. Those communities made large emotional and financial investments in community economic infrastructure, usually funded by state and federal grants in aid. Some of the proactive communities were more successful than others.

Frankly, most of the local governments that worked hard to reshape the community's trajectory were those where the pre-event trajectory was not particularly desirable: typically, the local economy was declining, or had not been developing at the pace or in the direction hoped for. Their recovery efforts were not particularly driven by community "boosterism" so much as by a deep concern for the future well-being of residents and an acknowledgment that without a successful intervention, the future was not bright. In many of these cases, local leaders hoped that the disaster would provide an influx of federal money and, thus, an opportunity to propel the community into a more favorable trajectory. Homestead, Florida, for example, obtained assistance from the federal government for several very large projects intended to help the community rebound from the loss of Homestead Air Force Base and from the effects of Hurricane Andrew. Montezuma, Georgia, received major assistance from both the state and federal governments in an attempt to transform it from a declining rural community into a retirement center and tourist attraction. The ambitiousness of the plans varied from place to place, reflecting what local leaders thought might be accomplished.

Not all the efforts were successful. Some cities, for example, focused their attention and a good share of the federal money they received on revitalizing their old, decaying, and largely vacant central business districts. Like city officials in many mid-sized communities, they were busy pursuing the nostalgic dream of a downtown teeming with shoppers and with exciting retail establishments, attractive civic activities, dining and entertainment establishments, and a host of clean, high-end businesses clustered in a handsomely landscaped urban setting. But despite their best intentions and efforts, that rarely happened. In truth, the central business district had been a losing cause before the disaster, most likely because another area and set of businesses, perhaps in a suburban mall, had taken its place to the satisfaction of community members.

RESTORING HIGH-PRIORITY STRUCTURES AND SERVICES

Recovery can begin with makeshift and make-do facilities and infrastructure. Developing a sustained recovery, however, requires that important buildings and infrastructure be repaired, rebuilt, or replaced. These activities often provide needed revenue for local retailers and service establishments, as workers who are brought in from other locations spend money on housing, food, entertainment, and perhaps even building supplies and equipment, thereby supplementing the revenues supplied by local residents. Moreover, even if the construction crews and firms are from outside the area, the rebuilding sends signals to potential investors that the community is, in fact, putting forth a serious effort to recover. And, of course, the new facilities will be more reliable and efficient than the makeshift and jerry-rigged temporary ones.

Local government can influence, albeit not control, outcomes and developmental processes. It can use public spending and public policy to "seed" the community with projects, facilities, and activities that will induce residents to invest desirable amounts time and money, thereby guiding or influence subsequent development. We are strong believers in both seeding and pricing to help shape the community.

REBUILDING OR RESTORING THE ECONOMY

Restoring utilities, rebuilding infrastructure, and repairing buildings is frankly a fairly simple process, assuming that sufficient money and technical skills are available. Restoring or rebuilding the community itself, however, is a much more difficult proposition, one with no guarantee of success. In this section, we identify and discuss a number of the challenges and suggest some directions that may prove useful.

Chickens and Eggs

Bringing workers back to the community makes little sense unless there are jobs waiting for them. Right after the extreme event, there is almost always a shortage of service workers in restaurants and lodging establishments and in demolition, cleanup, and construction. Some workers take on such jobs as temporary employment while looking for more suitable work. They need the income, and the shortage of workers often means higher wages for them. Long-term recovery will not occur, however, until stable employment is available. This poses a chicken-and-egg problem: employers may be willing to reopen only if there is an adequate workforce, and labor may return to the community only when jobs are immediately available. In these cases, the local government may be able to help find a solution.

One way that local government can help is to ensure the availability of a sufficient number of affordable housing units, which will make it easier for the community to attract new workers and for residents who left the disaster scene to return. In the short term, this means either providing temporary housing on site or bringing workers in from other communities in which they can find housing. Over the longer term, much more can be done in this regard. Florida City, for example, bought or was given lots, and then temporarily went into the home-building business, selling the homes it built at affordable prices. St. Peter, Minnesota built a subdivision for new housing.

It is probably unreasonable to assume that large numbers of workers will return to a badly damaged community until schools and other family-oriented facilities are back in operation. While we usually expect those who evacuated the community to return, they do not always come back, especially if they have found a better life in the community to which they found shelter. They need one or more good reasons to return.

Working to Revitalize Local Business

Some local governments help businesses get back on their feet by targeting those businesses that are essential to the community and facilitating their permit applications and health inspections. Most mid-size and smaller communities we studied tried to provide modest help to other small business owners as well. But many others did little or nothing to assist local businesses with their financial needs arising from the disaster. In fact, most had very little experience in this regard. Local governments are more likely to have experience trying to attract new businesses and industry to the community, so that is what they did. Too often, when confronted with problems, we do what we know how to do rather than what needs to be done.

Sometimes the actions that local governments take to help local businesses recover are actually counterproductive. In Northridge, the owner of a camera store at the heart of the damaged area reopened his shop and was trying to accommodate the residents' demand for cameras and film to document their losses. His street was blocked off by a National Guard cordon, so he attached a small sign to a nearby cyclone fence to help people find his shop. The city of Los Angeles ticketed him and fined him $300 for posting an illegal sign. For years, he had the same bitter complaint: "Here I am in the middle of a huge disaster zone, trying to stay in business, generating sales tax revenue for them, they've made it impossible for people to get to my store, and then they have the gall to fine me!"

Where the local governments we studied did provide help to small businesses, such help was usually funded, in one way or another, through a combination of Housing and Urban Development Community Development Block Grants, private donations, and actions taken by local financial institutions. Sometimes the help was the result of the local chamber of commerce or a local banker advocating for help to small businesses; in those cases, the assistance almost always took the form of small loans. The loan programs usually offered some kind of subsidy, including low or no interest, deferred repayment, and possible forgiveness of the principal, and local banks sometimes subsidized the interest rates themselves. At the same time, these loans always had some strings attached. Usually, the business had to agree to do business within the local jurisdiction for some number of years to earn loan forgiveness. But such requirements can become particularly burdensome: for example, it rarely makes sense to stay in the same business location if no one else is there. The local jurisdiction may want to revitalize its fading central business district and require borrowers to do business there, whereas the owner just wants to start making money again.

The Small Business Association (SBA) is authorized to make loans to homeowners, nonprofit organizations, and businesses, regardless of size, in areas covered by a Presidential Disaster Declaration. Three kinds of loans are available: physical disaster home loans, physical disaster business loans, and economic injury business loans. The loans are made at lower interest rates than typical of private financial institutions. SBA loan criteria include "reasonable assurance that you can repay your loan out of your personal or business cash flow, and you must have satisfactory credit and character."[34] SBA practices may vary somewhat from disaster to disaster: "Since SBA cannot predict the occurrence or magnitude of disasters, it reserves the right to change the rules in [the disaster loan program] without advance notice, by publishing interim emergency regulations in the Federal Register."[35]

Despite significant investments made by the SBA in disaster areas in the form of loans to homeowners and businesses, some of the business owners we interviewed had complaints about its loan programs. Since the loans are made primarily on the basis of the business before the extreme event, and because business conditions after the event are often quite different from what they had been, the loans became an albatross for business owners instead of the hoped-for solution to their problems. The SBA requires that disaster loans be collateralized to the extent possible by the borrower, which often requires the borrower to use his or her home as collateral. Several small-business owners complained that the loan program made it virtually impossible for them to make necessary changes to their businesses in order to adapt to new circumstances. Some people claimed that the SBA had "sold their loans" to private firms and that it was impossible for them to get partial releases of collateral pledged for their loans. Although SBA officials have told us that the SBA does not sell its loans, we did not investigate the claims of the people we interviewed to fully understand what problems they were encountering. Rather, we were concerned with people's perceptions, regardless of their accuracy, since it is perceptions that drive people's behavior.

Local governments can do many things to help local businesses recover and to facilitate the recovery of others. Not all parts of the infrastructure and not all services have to be resumed immediately. Along with placing a high priority on critical utilities, infrastructure, and services, it is critically important to get the most urgently needed businesses open first. Some businesses need loans to get back on their feet; others do not, and still others may be ineligible for financial assistance from state and federal agencies, depending on their financial history. A small business may not even be eligible for a loan from the SBA. Thus, local governments that want to help jump-start recovery efforts need to talk with local business owners and managers to learn what it is they need urgently to meet the high priority needs of their customers.

CLOSING THE LONG-TERM REVENUE GAP

It is a big challenge to close the long-term post-disaster revenue gap, particularly if much of the community's tax base has eroded and businesses have closed or moved away. In the latter situation, the answer is often economic development. It is crucial to have a strategy to guide economic development activities, if for no other reason than to be able to use federal and state monies effectively.

The strategy should begin with a rigorous, honest assessment of what kind of development is possible and most congruent with the community and its workforce, location, and resources. We caution against making these assessments in isolation. The decisions and consequences are so important that it makes sense to get a variety of perspectives and to guard against "groupthink." Business owners and managers running local firms may be able to provide some insights about the kinds of businesses and firms that would complement their own and that might be created or induced to build in the community. Some state governments employ economic development specialists who can help.

Once decisions are made about the kinds of industry or businesses might be induced to come to the community (a good option) or created from within the community (often a better option), the next step is to determine which approaches might be most effective in bringing that about. Economic development planning might be done autonomously by local officials, but when stakes are high, it is extremely important to ensure that the vision is practical as well as grand and that the strategy is likely to be effective. To this end, competent professional advice and counsel may help to ensure that objectives are achievable.

Presenting plans to a wide range of constituents for evaluation before they are scheduled for adoption and implementation is one way to assess the viability of the strategy. Of course, such a tactic can also bring to the fore a high degree of conflict, so officials will want to make sure that individuals who are highly skilled in conflict management techniques are involved in any public vetting of ideas and plans. Otherwise, that which was intended to enable comprehensive evaluation and a sense of ownership can yield dysfunction that drags out the course of recovery.

States have a role to play as well in helping to close a local government's long-term revenue gap in the aftermath of an extreme event. Those states in which local governments are "held harmless" for the effect of disasters on shared taxes and state grants to local governments do a great service to their communities. Some things may be relatively easy to accomplish before a disaster, such as modifying formula-based grants and shared revenues so that a sudden drop in population caused by a disaster will not penalize the community. More complicated legislation is required after a disaster when, in an effort to backstop local government finances, the state seeks to compensate the local government for income it might have had, had not the event occurred. In the first few years of rebuilding southern Dade County after Hurricane Andrew, the state of Florida estimated the amount of sales tax revenue generated in the rest of the state that presumably would have been spent in the ravaged areas, and it provided those funds to local governments there. But that kind of calculation is often difficult.

The length of time that a local government may need assistance will vary, depending on the nature of the catastrophe and the local community's ability to rebound. States should take steps to ensure that special tax and revenue treatment lasts long enough to get their communities through the roughest parts of recovery. There is, of course, no replacement for rebuilding the community tax base. For some local governments, this will require their continuing front-burner attention.

Finally, state governments can help by requiring that local governments participate in some form of disaster insurance program. Creating a reserve for disaster recovery makes life easier for budgeting at all levels of government.

HELPING WITH HEALING

Most communities do something to commemorate a disaster. Many hold events on the first or fifth anniversary of the disaster, using these occasions to mourn the dead, recite the trials and tribulations that residents have suffered, and extol the progress that has been made toward recovery. Such events usually contribute to a group catharsis and to community bonding around the shared experience. Sometimes memorials are built. Grand Forks, North Dakota, erected a sculpture and put up permanent signs identifying what stood on various sites before the flood. One Gulf Coast community that suffered through Katrina created a sculpture of found objects and debris from residents' homes, and put it up in a small park designed for contemplation and reflection. Most communities have created some kind of permanent marker.

A few communities provided mental health services to those who had experienced the disaster and were having emotional or psychological difficulties associated with it. We most often heard respondents talk about experiencing chronic and acute depression, although the literature identifies post-traumatic stress disorder as an effect.[36] We studied communities where the local government provided assistance for up to six months after the disaster although the symptoms sometimes did not manifest themselves until after that.

Previously we noted the efforts of a nonprofit organization in Tarboro, North Carolina, to help children deal with their disaster experiences, but we have not seen evidence that such help is the norm in other communities. FEMA now provides assistance to communities for mental health services following disasters, but application for the assistance must be made relatively soon after the event and must provide evidence that the need exceeds the capacity of the state and local government to provide the services.

When we look at individual and community healing after a disaster, we see both immediate and long-term needs. In the immediate aftermath, we see a need for comforting, for assurance and reassurance, and for medical, psychological, and psychiatric care. Over the longer term, we see a need for community catharsis and for symbols of hope and recovery. Healing is a process that takes some people more time than others, so the services are needed for an extended period after the event. Healing is also a collaborative process, facilitated by the understanding that one is not alone.

Chapter 17

BASIC STRATEGIES FOR REBUILDING THE LOCAL ECONOMY

"Wishing, of all employments, is the worst."

Edward Young

After Hurricane Katrina swamped Liuzza's in 2005, its owners worked to reopen the neighborhood bar and restaurant as quickly as possible. This photograph, taken in March 2007, shows a restored Liuzza's.

NOT ALL COMMUNITY ECONOMIES UNRAVEL after an extreme event. Unfortunately—and particularly for communities not favored with a great location, desirable amenities, or the other magnets that attract economic activity—extreme events often create very serious economic problems for months, years, and even decades into the future.

The challenges of establishing or reestablishing a viable economy that could be sustained over an extended period are enormous: across the country and internationally, everyone else wants the same thing and is competing for it. A significant part of the challenge is that the local economy is truly self-organizing and, to a very considerable extent, outside the ability of local officials to influence, much less manage, it. Individuals and businesses make choices about what to do and where to do it. In communities that are off the beaten path and that have lost their locational or amenity advantages, locals are faced with the extremely difficult challenge of creating a new raison d'être for the place—one that is strong enough to attract development.

Instead of, or in addition to, helping individual businesses, local governments sometimes try to use the disaster as the launch pad for making long-term improvements in the community economy, using available state and federal money not only to rebuild but also to undertake economic development projects. This is most likely to happen in communities that were having economic problems before the extreme event.

SEVEN STRATEGIES

We have found that local governments, collectively, selected from among the following seven basic strategies in their efforts to foster post-disaster economic development.

Do What We Already Know How To Do

One popular approach is for the local government to continue to do what it was doing before the event. This usually means attempting to attract business and industry by (1) advertising the attractiveness of the local community, its location, local amenities, and the sterling workforce, and (2) offering incentives such as free land and/or buildings, tax abatements, and the like. This approach is widely used, but its relative ineffectiveness is demonstrated by the massive number of largely vacant municipal industrial parks across the United States.

The problem is that this approach usually attracts mobile industries—industries that can pick up and move quickly and that are likely to leave town when the incentives run out. We used to call it "chasing smokestacks" or "chasing shoe factories." While people no longer want smokestacks and very few shoe factories exist in the United States, there are footloose industries that tend to move from place to place in search of tax breaks, low taxes, and low-cost labor. If a community is in trouble, these industries are a lot better than nothing, but usually not for long.

Re-Create What Used To Be

Many local officials seem to hold the passionate conviction that there should be a central business district in their communities where people shop, do business, and gather. Historically, the core of the city was where one found the cathedral, the castle, the banks, the markets, and the musicians. Despite the fact that central business districts in mid-sized cities are now largely historical artifacts, having been replaced by suburban malls that follow sprawl, many local governments still try to bring them back.

For example, although Grand Forks, North Dakota, and East Grand Forks, Minnesota, did a mostly superior job of recovering and accomplished much that other places might well try to emulate, they each spent a large amount of federal grant money trying to recreate central business districts that were moribund before the flood. Retail businesses had been moving away from the downtown area and toward the edge of town, where the people lived. In the case of Grand Forks, enormous sums were spent to rebuild and revive something that isn't working much better now. Across the river in East Grand Forks, the city helped move a popular restaurant and bar a couple of blocks north and about twenty feet higher. That worked out well because the move made the successful business more attractive. However, when the city bought a small downtown shopping center and converted it into a place that it could lease out to small businesses, that didn't work out quite as well because the customers still went primarily to the new, auto-accessible shopping areas on the city's periphery.

Actually, using "free money" from the federal government in an attempt to recreate a vibrant central district makes more sense than trying to use money scraped from the local tax base. In some ways, it is a no-cost approach to addressing a project that has considerable appeal but a low probability of success. At the same time, though, there are opportunity costs, and those monies might have been better spent on addressing or remedying other local concerns or problems.

After the fire in 2000, Los Alamos, New Mexico, also tried to revitalize its central business district—an area that had been suffering economically before the Cerro Grande fire. In addition to the normal challenges of maintaining a central shopping area, Los Alamos faced two additional challenges. First, Los Alamos is basically a one-business town: it depends on the Los Alamos National Laboratory. The laboratory had recently gone to a four-day week and had built a new cafeteria onsite for its workers. Second, there was a newly improved highway down the mountain to Santa Fe, which brought that city a lot closer to Los Alamos, at least in terms of travel time. Thus, whereas laboratory employees had previously eaten lunch and shopped mostly in downtown Los Alamos, they now ate at the Lab's new cafeteria on their work days and went to Santa Fe to do most of their shopping on their long weekends. The city used the fire as a lever to try to revitalize downtown businesses, but it couldn't change the proximity to Santa Fe or the food at the lab facilities.

Take a Great Leap Forward

Of all the cities we studied, we were most impressed with Homestead, Florida, and its bold approach to economic recovery after its triple disaster. Hurricane Andrew essentially flattened the city, driving out many of its residents. The closing of Homestead Air Force base drove out many more residents. Finally, at least locally, NAFTA is thought to have had a major adverse impact on the dominant agricultural economy in the area. Local officials thought hard about how to use funds from the federal Economic Development Administration to launch Homestead into a new economic orbit. The strategy they came up with seemed sound, but almost fifteen years later, despite massive amounts of federal assistance, it has not worked out as hoped.

The city focused on creating attractive facilities and amenities to bring business and retirees to the community. Just before Hurricane Andrew struck, the city had completed one such project: an $18 million spring training center for a major league baseball team, the Cleveland Indians. The Indians had decided to move to Florida after a half century of conducting spring training in Arizona. After the hurricane essentially destroyed the facility, the city rebuilt it quickly, using federal grants. The park was opened in 1993 and refurbished in 1996, but the Cleveland Indians had already moved to Chain of Lakes Park near Orlando and were not about to move again. Homestead is now the owner of a large baseball facility that doesn't see much baseball. The paint has begun to fade and peel.

A second major project was the Homestead-Miami Speedway, which was intended to cash in on the popularity of NASCAR. The track is doing fairly well, hosting a solid program of racing. Unfortunately, a law suit initiated after the city had developed the track resulted in a Florida Supreme Court decision that resulted in the city having to pay *ad valorem* property taxes to Miami–Dade County on the assessed value of the facility. That created an unexpected and significant drain on the city's finances just when local officials were looking forward to a significant cash infusion to both the community and the city treasury.

The third effort involved the development of a very attractive 280-acre industrial park complete with infrastructure. Local leaders had hoped to convert the former Homestead Air Force Base into a commercial airport to facilitate trade with Latin America. The park was developed in collaboration with the Rockefeller Group Development Corporation. For reasons we were unable to fully uncover, the plans to use the airport fell through, and the old base now serves as home for air reservists. The commercial airport had been intended to be pivotal in the area's economic development. Whether it was the fact that the city could not obtain rights to the airport or some other combination of factors, the industrial park remains largely vacant fifteen years later.

Ironically, after all this effort and money was invested in more than a decade of unrealized hopes for revitalization, Homestead is now growing. Miami is expanding southward, and Homestead is the next available area for housing. As a result, people from Miami are turning Homestead into a bedroom community as thousands of new homes have been constructed there since 2003. This has generated major increases in the local tax revenue, but it has done little to improve local employment options. Essentially, it reflects a fundamental change in Homestead's historic role in South Dade County, as well as a new trajectory for the city.

Transform Yourself

A few cities have tried to transform themselves after a disaster. One story is particularly illustrative. Montezuma, Georgia, a small town of about 3,500 residents on the Flint River, is located about fifty miles south-southwest of Macon and about sixteen miles from the nearest access to Interstate 75. Montezuma was created when a railroad was built to cross the Flint River at that location and to serve the surrounding agricultural area. But rail use has declined as the agricultural economy has changed, and because of the improved highway access to larger towns those larger towns have increased their market share at Montezuma's expense.

Local officials have struggled to find ways to make Montezuma attractive to both industry and retirees, but without great success. The local government continually works to attract employers to the area, and low-cost labor is available; however, that labor is mostly unskilled and undereducated, and many locations provide better access to the interstate highway system. Local merchants had originally relied on local trade because they were conveniently close and did not face competition from national and retail chain stores. As that competition emerged and became strong, most merchants remained in business by providing exceptional customer service or by extending credit to those who might otherwise have difficulty obtaining it. Even so, with good highways leading to Perry, twenty-five miles away, and to larger cities farther north on the interstate toward Macon and Atlanta, large chain stores have increasingly drained away local business.

Montezuma flooded in July 1994 as a consequence of heavy, persistent rains produced by tropical storm Alberto. Small earthen dams creating farm ponds along Beaver Creek failed under the deluge, and water ran overland into downtown Montezuma. The old core of Montezuma, dating to the 1850s, was built in "a low swamp in the midst of a dense thicket of woods," where the railroad crossed the Flint River. Being in the lowest part of the town, the old business core had also borne the brunt of floods in 1902, 1929, and 1948.[37]

After Beaver Creek overflowed, business owners watched as the Flint River rose and overtopped the levee built by the Army Corps of Engineers in 1956. The river reached its highest crest in recorded history—about thirty-five feet. The water, muddy and colored by the red clay typical of Georgia, remained in the business district for six days, trapped by the levee as the flood swept downstream.[38] The corps refused to permit city officials to breach the levee to drain the downtown, but local lore has it that a local official drained the town center by cutting a hole in the levee with a backhoe under cover of darkness. In any event, the downtown drained, leaving the entire urban business center muddy.

City officials tried to look on the bright side: the flood was an opportunity to get help and resources to rebuild the picturesque old business core, which was generally neglected, and perhaps reestablish it as a place to go, to shop, and to attract retirees and tourists. The community is close to Andersonville, site of the infamous Civil War prison and home of the National Prisoner of War Museum, which draws a relatively small but reliable number of visitors annually. The plan was relatively simple.

The city worked with state representatives, who helped to rehabilitate the business district into a quaint replica of what it had been many years before. With help from the Georgia Historic Preservation Division, the city acquired title to the façades of the commercial buildings in the old central commercial core and then repaired them, painted them in complementary colors, added attractive street architecture, installed old-style street lamps, replaced pavement, erected statues and monuments, planted trees, and created cozy open spaces. Thus was the downtown transformed into a pleasant place for shopping and visiting.

Unfortunately, all this effort never paid off. The rejuvenated business district did not generate much in the way of new employment, the tourists never came in the numbers hoped for, and neither did well-off retirees. By 2004, many of the downtown merchants were barely hanging on. Some had closed and left their buildings vacant or were replaced by marginal establishments aimed at lower-end buyers. The wonderful paint jobs are peeling and the dream is fading.

The Montezuma effort did not attain the hoped-for long-term community resurgence, but not because the local officials did not develop a thoughtful plan for transforming the community. The city was hampered by considerable poverty in the area, a largely uneducated workforce, few employment opportunities for locals, and national trends that make it tough for small towns to survive. It was not able to develop a big enough magnet for tourism, shopping, and retirement living to overcome the basic problems of the community and the immediate region. To further compound the problem, the construction projects took more than a year, during which access to downtown and its shops was extremely limited and inconvenient. Consumers change habits when confronted with "hassle costs" while trying to do what they had been doing.

Work With the Market

Florida City, immediately adjacent to and smaller than Homestead, was also devastated by Hurricane Andrew. Before the hurricane, Florida City was a primarily African-American community with relatively few employment opportunities and about 40 percent of the households living below the poverty line. After Andrew, it stood largely in the background as assistance poured into Homestead; however, Florida City had two advantages over Homestead. The first was location. Within Florida City's boundaries lies the intersection of the Florida Turnpike and U.S. Highway 1. Second, the city was very stable politically. The mayor had been in office for many years, as had several members of the small city council.

After the hurricane, large retailers began visiting the area because U.S. Highway 1 is one of only two routes to the Florida Keys, and there are virtually no large building sites between Florida City and Key Largo, thirty miles to the south. The mayor and his staff assessed the situation and made a prudent decision: Florida City would simplify approval processes when a firm thought it might want to locate along the highway near the toll road terminus. The mayor and two or three other city officials would meet anytime, anywhere, with prospective businesses on behalf of the city; they would be able to "cut the deal" with the private firms in a single meeting, and the decision would stick. Rather than having to appear before five or more committees over a series of weeks or months, the business representatives could meet once and be assured that the deal would be approved. We are inclined to think that Florida City was able to implement that strategy because it had a longtime mayor and stable city council. If local politics are factious, local officials do not tend to enjoy long incumbencies. Florida City was thus able to empower the mayor and top local officials to focus on reducing the "hassle factor" for businesses. It worked.

Between its locational advantage and its streamlined approval processes, Florida City attracted two "big box" retailers within two years of the hurricane. The new buildings augmented a small tax base, and the box stores provided many jobs. Although the jobs were mostly at minimum wage, they afforded employment to many who had little employment history, few skills, and limited mobility.

Build on What You Have

Three communities we studied provide good examples of bringing about economic recovery by building on the assets that the community had before the disaster and, for the most part, still had afterwards. St. Peter, Minnesota, is a community of 10,000 people located in the east south-central part of the state. Founded in 1853, it is one of Minnesota's oldest cities. It is surrounded by rich farmland and serves as an agricultural-business center; it is also home to Gustavus Adolphus, a small liberal arts college, and a state hospital. Minnesota Highway 169, which runs through St. Peter, connects it with Minneapolis–St. Paul, about an hour away, and with Mankato, twenty minutes to the south.

On Sunday, March 29, 1998, a tornado caused severe damage to the community. The tornado had a particularly wide swath, up to a mile and a half wide in some parts. It damaged or destroyed 40 percent of the homes in the community, many buildings in St. Peter's downtown commercial area, and every building at Gustavus Adolphus College. Fortunately, Easter break at the college began the day before, so the campus was essentially vacant when the tornado struck, and no one was killed or injured there. Because the warning sirens worked, only two people in St. Peter were killed, despite the ferocity of the storm. But the town was without water, sewer, or electricity for nearly a week, and damages were estimated to be $300–$500 million.

Funds from FEMA helped considerably with covering the cost of rebuilding and repairing infrastructure. The city was insured against tornado damage, and its insurance company gave it a $1 million "down payment" to start repairs. The fact that many of the property owners were insured against wind damage greatly aided the repair of privately owned property. (Unlike the case with hurricanes, tornadoes usually lead to relatively little conflict between policy holders and insurers about what damages were caused by wind and what were caused by water.) Gustavus Adolphus College was insured as well, but it also received considerable financial assistance from its alumni, which facilitated rebuilding and improving the campus. And students helped a great deal with cleanup, both at the college and in the community.

St. Peter suffered mostly immediate consequences but relatively few systemic community consequences. Neither the economic nor the social fabric of the community was severely damaged. Furthermore, much of what the community achieved in the months and years after the tornado was already in the works or in the planning stages before the tornado struck. St. Peter's Comprehensive Plan, developed three years before the tornado, had roots in a document prepared for the chamber of commerce twenty years earlier. The introduction to the 1995 plan states:

> This comprehensive plan can provide a framework for continuing the story, setting the most general of limits, laying down some rules, ground rules, so to speak. It offers a suggestion here, chosen direction there, and organizes the continuing conversation because in one sense the plan is never "finished." *It serves as a friendly guide; follow it but make adjustments depending on the flow of the shaping forces and events* (italics added).[39]

Local officials did not hesitate, nor did they wait for help from a federal agency. Although the tornado shaped some aspects of St. Peter's future, the existing plan and vision was the touchstone of the recovery process. In the following spring, the city hosted a three-day visit from the Minnesota Design Team (MDT), a volunteer group of architects, landscape architects, urban designers, planners, and other experts in design and community development that has been helping Minnesota communities develop shared visions for improving their physical and environmental design since 1983.

As a precursor to the MDT visit, St. Peter's officials were asked to cite the three most important problems that the community faced. They identified recovering from the tornado, defining a common vision for the future, and acquiring resources to meet that vision. When asked to anticipate the problems the city would face in ten years, they responded with a short list:

1. What do we do with an older downtown area that is changing from storefront to service-based business?
2. How do we balance economic growth and the need for housing?
3. What can we do to make our landscape and river an asset to our community?

When asked to "describe the three best opportunities for your community today," their response was

1. Our ability to make changes with deliberate forethought after the tornado.
2. A great spirit of cooperation within the community and its groups.
3. The chance to put new infrastructure in the ground that may affect our community's long-term future.

The early emphasis of St. Peter's recovery process, like that of most places, was on the repair and reconstruction of municipal facilities, utilities, and infrastructure. Fortunately, a significant part of the cost for a new library had already been willed to the city by a benefactor, so a new library would have been built even if the old one hadn't been destroyed. The community center was also destroyed by the storm, but a new facility had already been planned and the early insurance payment (along with a grant and additional city borrowing) advanced the construction schedule. On the other hand, a municipal affordable-housing project was delayed a year or more because St. Peter's priorities became reordered. Upgrading the city's electrical power generation plant was in the planning stages before the tornado, and after five days without power, the upgrade project gathered new support. The city also added a fiber-optic capability to its infrastructure and moved power and phone lines underground.

The municipal government, the nearby Minnesota state hospital, the local college, and the people of St. Peter established a level of collaboration and interaction that has continued since the storm almost a decade ago. In the 2004 Comprehensive Plan, local officials looked back at the recovery period:

> In the aftermath of the 1998 tornado many previously held ideas and thoughts about our community were brought once more to the forefront, to be weighed against the current reality and our wishes for the next 20 to 30 years. These ideas were challenged based on the hope that the tornado, which brought tremendous devastation, also brought a clean plate, that golden opportunity to rethink the past thoughts and make sure they match our current hopes.[40]

Biloxi and Gulfport, Mississippi, also appear to have taken appropriate steps to rebuild their local economies after Hurricanes Katrina and Rita by building on what they had before and what remained. Locals tell us that the storm surge swept ashore at twenty-two to twenty-four feet in height and with considerable velocity. The first three to five blocks inland from the shore were essentially wiped clean. Antebellum homes were swept away; Jefferson Davis's shoreline home was severely damaged. Biloxi was hit extra hard because the storm surge continued around the east end of town and swept in from what is called the "Back Bay."

Biloxi is home to Keesler Air Force Base. The second largest medical center in the Air Force, Keesler provides medical care for approximately 75,000 eligible beneficiaries in the Gulf Coast area. Apart from that, beautiful sandy beaches, warm weather, and casino gambling mean that tourism and retirement living drive the local economy. The beaches didn't go away after the hurricanes, but the cities had to clean up enormous amounts of refuse and debris from the beaches and the water, and they had to put devastated public utilities and buildings back in place so that services could resume. Efforts were made to ensure that the replacements would be more robust against future hurricanes. Biloxi, in particular, took pains to exercise land use planning and controls that would put housing on safer ground. A considerable amount of prime land along the shore was set aside for recreational purposes.

The casino owners rebuilt fairly quickly. Eighteen months after the storm, some were already in back in business. Chain hotels and restaurants rebuilt at least some facilities. The shore is largely devoid of private housing right now; there are still hundreds of FEMA trailers. Although the need for hotels and restaurants is not yet as strong as it was before the hurricanes, it is growing. Biloxi and Gulfport are taking the resources they have and building on them. These resources are hard to duplicate elsewhere in the country. Between hurricanes, at least, they provide employment and a tax base.

Don't Do Much Other Than Restore Infrastructure

Los Angeles adopted the "don't do much" strategy for business recovery after the Northridge earthquake. The city and state worked quickly to repair the badly damaged Santa Monica Freeway and restore utilities to the damaged areas. But apart from providing financial support, the city left to nonprofit groups any efforts to help local businesses get reestablished. From the city's standpoint, the strategy worked: people are employed and there are business-es in the damaged areas. But in fact, many businesses failed in the aftermath because their circumstances had changed so much, and most were replaced by other businesses owned by other investors.

Los Angeles's approach is understandable in its context. Although dam-age was extensive, it did not cripple the metropolitan area. It did not seriously threaten the regional economy. Nor did it result in a substantial blow to local governments' property tax base. And even though many individual business owners lost their businesses and their assets, others were willing and able to fill in for them, so residents were able to get what they needed when they needed it, albeit from new sources. Moreover, while we do not know whether it is the case in Los Angeles, many governments do not feel it is either impor-tant or appropriate for them to help private organizations recover from disas-ters. In this view, they simply reflect federal policy.

WHAT WORKS?

Everyone knows that a salesperson's life is a lot easier when he or she is selling what everyone wants. Communities with a great location, super ame-nities, low crime rates, great schools, and fine housing at low cost will not have much trouble with economic development. On the other hand, few com-munities have all that. Worse yet, a good location can turn into a not-so-good location as markets shift, tastes change, and new products and technologies replace old ones. For the communities that do not have everything going for them and that also suffer an extreme event, the prospects of developing a via-ble economic base to support a viable community can be bleak.

If we knew what to do to create strong economies in places where they are lacking, we would probably be busy doing that: working, charging high fees, and setting up franchises around the world. We have, however, learned some things that seem to help. For example, it makes lots of sense to evaluate a community as realistically as possible to learn what competitive advantages, if any, it might have for one or another economic base. Sometimes it will be difficult to come up with more than a dream of what might be "if only. . . ." Other times it might require bold strokes to move in considerably different directions. The key appears to be envisaging what might be both possible and desirable and then working to facilitate the transformation.

Moving in a bold new direction sometimes works, too. In the late 1880s, Wisconsin was one of the nation's primary wheat producers. However, disease destroyed the crops, Minnesota and the Dakotas opened as farming areas, and Wisconsin was left without much of an economic base. One of the University of Wisconsin's regents believed that the state could convert its farms to dairy and cheese production. The idea was unpopular among farmers: milking was women's work. However, the university's Department of Agriculture went to work. It introduced year-round milking, silos and silage, dairy cooperatives, agricultural short courses, and a host of inventions to facilitate producing and marketing milk and creating cheese. By 1910, the agricultural transformation was essentially complete. It is a remarkable story of how government can "seed" a massive economic transformation.[41]

It also seems to make considerable sense to work hard to retain the most desirable elements of the community's economic base following a disaster. Helping an existing industry or business is probably significantly more productive than trying to attract a new employer from somewhere else or to create a new economic base.

CHAPTER 18

CREATING AND IMPLEMENTING A STRATEGIC PLAN FOR LONG-TERM RECOVERY

"In the middle of difficulty lies opportunity."

Albert Einstein

An example of a Make It Right home in New Orleans's Lower 9th Ward. Make It Right is a program established by actor Brad Pitt. Photograph taken in New Orleans in September 2008.

*I*N THIS CHAPTER WE IDENTIFY some approaches to recovery planning and some characteristics of recovery plans that we have come to believe have a higher probability of success than others. These are lessons we learned in the communities we studied; some are from good examples and some are from not very good examples.

BASIC CONCEPTS

We have already discussed that community recovery is neither assured nor automatic. If you rebuild it, they might not come back. The challenge is to think strategically about how to facilitate and encourage recovery of the complex, ever-changing community system. It is essential to know that not every outcome can be controlled and that random events will continue to thwart this or that plan. Flexibility, adaptability, and a tolerance for ambiguity are essential.

Systems Thinking

Viable communities are those that work well as self-organizing systems adapting effectively to changing circumstances. Trying to jump-start a community that has suffered massive losses from an extreme event and in which cascading consequences have destroyed its economic life and social order invites unwanted images of trying to animate Frankenstein's monster with massive jolts of high voltage. Nonetheless, it is important to think of the community as a system. Some sparks are needed after the extreme event to trigger positive responses from those who remain and those who hope to return. Government and community leaders can provide the signals and seeds to create those sparks and to nurture them.

One of the key elements of systems thinking is understanding that all the parts are interconnected and have to develop roughly in tandem with the other parts. Residents need groceries, and grocery stores need customers. Businesses need employees, and employees need jobs. Some things, of course, must come first. Everyone needs electricity, sewer, and water very early on. Other things can develop in tandem.

Coping with Uncertainty and Ignorance: Policies as Guides for Action

During recovery, decision makers will be faced with scores of important choices. They will also be faced with uncertainty about how some things are unfolding and with ignorance of other phenomena yet to manifest themselves. How a particular question is decided will, almost invariably, have consequences for future decisions. At the same time, it is completely unreasonable to forestall making all the decisions that have to be made. It is therefore important to develop policies that can serve as criteria for helping to evaluate the alternatives. We caution against making decisions before they are needed: for example, it isn't necessary to decide on a location for the new library in town while the city is still under water. But it is important is to define some criteria or policies that will be used to evaluate sites for the library—perhaps on a main bus line or near the center of town—when the time comes to make that decision.

Because making firm decisions limits future options, it sometimes makes sense to take what we like to call a "soft path." Soft paths are solutions that are revocable. They sometimes involve using operating funds rather than capital project funds. Uncertainties can be resolved or, sometimes, may even resolve themselves. If, when faced with ignorance, there is a way to use a soft path, it is often better than trying to change things once the concrete has hardened.

The Principle of Incompleteness of Design

Communities, as we have said, change every day. Any community plan is just a snapshot of what the community might look like at some time in the future. The city will never be finished, and so it has to be designed to facilitate change and to accommodate the unexpected event and the emergent need or desire.

COMMUNICATION

It is important to communicate with the full range of constituencies in the community and beyond, including local elected officials, department heads and employees, community residents, homeowners, business owners, state and federal officials, and media representatives. Some constituents will require more communication than others, but the messages must be consistent.

One of the most critical plans that must be developed extremely early in the recovery period is a community communication plan. Every community we studied had some way to communicate with the public during the emergency period, and each one tried to keep the public informed about what was going on in the city during recovery. Only a few, however, had plans and procedures for doing it effectively. The following observations and suggestions stem from our conversations in those communities.

Functions of Communication

Formal communication from local government to its constituencies serves both symbolic and substantive purposes. It is important for citizens to know that the local government is on top of things: operating, aware of what is going on, and doing what needs to be done. It is similarly important to know that there is a source of accurate, reliable information. When everything else is in disorder and chaos, people need anchorage. Some of that anchorage may be provided by the local media, but after the emergency period, most information that the public receives must ultimately come from the local government itself.

The substantive elements of communication from local government after a disaster will change over time. Initially, the content is focused on what happened, where to get help, how to make contact with friends and family, when and where meetings will be held, where to buy gasoline and groceries or get medication, and what stores and restaurants are open. Later, the focus typically shifts to providing information needed by those working on their own recovery challenges; such information might include actions that the local government is taking; schedules for debris removal and trash collection; the date for resumption of mail distribution; and timetables for the expected completion of various government tasks, such as the restoration of power, water, and sewage treatment, and the reopening of the bridges and airports.

An important function of communication from city hall is ensuring that the information it disseminates is seen as credible. There are essentially three steps in this process. The first step in ensuring information credibility is making sure that it is truly accurate, reliable, and accessible. The second step is making sure that it comes from a credible source, one that is trusted. Rumors are always rampant during and following disasters. Neighbors talk with neighbors, and facts or half-truths become distorted—almost always for the worst. It is like the telephone game in which one person in a circle whispers something to the next person, who whispers it to the next person, and so on, until the message has gone completely around the circle and is revealed to have been substantially altered in the process. Imagine that writ large and spreading, not in a circle, but in a growing wave across the community and then echoing back, badly distorted. Overcoming rumors is hard work and not always effective. Providing timely, accurate, reliable, and accessible information at the outset can help nip them in the bud. Similarly, local government officials need to remember that reporters do not always get it right, either. After a disaster, reporters are likely to be pulled off other beats, like sports, to cover the story. Providing written press releases is more likely to result in accurate reporting than depending on reporters who often have limited knowledge of the natural phenomenon and even less knowledge of how local government works.

The third step in ensuring that the information issued by city hall is seen as credible is making sure that any missteps are immediately recognized and addressed. Mistakes are inevitable; addressing them requires courage and diligence.

Typically, only larger local governments have staff members who are trained to communicate effectively with the outside world or whose primary job is to provide public information. For those that don't, the only way to ensure that messages are consistent and accurate is to have a policy on who is to talk with whom about what. Local officials should designate a principal contact point and explain the communication policy to all the staff. Depending on the circumstances, it might make sense to have several contact points, each with a designated area about which they are able to talk authoritatively. It may also be advisable to hire a communications consultant to help establish communication policies and practices, and to help with technical matters.

Two-Way Communication

Communication has to flow from the city and to the city. Too often, we have seen situations in which local officials set out to help one or another group without ever talking with members of that group. For example, most local government administrators know little or nothing about small business, yet in city after city they sat in their offices, imagined what local small business people needed, and then devised programs that were either not useful or downright dysfunctional. Before they assume that something is needed or not needed, local officials need to talk with those who are going to be affected to learn what is likely to be effective or detrimental.

Similarly, a local government must communicate simply and in a timely fashion with community members who need information about the local government's plans and timetables before they invest money in rebuilding. We recognize that this may be easier said than done. Still, community members should not have to listen to dysfunctional rhetoric (e.g., "We're going to rebuild New Orleans just as it was") in place of concrete plans. Rhetoric ultimately angers people, especially when it contains empty promises. Eventually, many people will take their money and leave the dysfunctional community, exacerbating the recovery situation.

Choice of Delivery Systems

It is important to set up communication processes before the disaster. If local officials wait until after something happens, mistakes will be made, time will be lost, and information will not get to those who need it. People should know where to look for information before an extreme event; they shouldn't have to go hunting after the fact. Doing so only adds to their uncertainty and anxiety.

There are many options for getting needed information to people on a timely basis. Some school districts have automated telephone systems that advise parents that school will be canceled or delayed because of a blizzard, ice storm, or threat of a severe storm. After disasters, some local governments set up a Web site where people can go to get up-to-date, reliable information. If this site is not part of the local government's official Web site, the official Web site will provide a hyperlink to a site that is dedicated entirely to community disaster and recovery news. Hospital officials who we interviewed told us that it is often useful to have the Web site generated in another community so that if power is disrupted in the local community, the site is still operable. Obviously, if the power in the local community is disrupted, those who are there will not be able to access a Web site, no matter where it is housed. Experience suggests, however, that people who have evacuated will rely on such sites for information about when to return and what they are returning to. For those who remain in the community, simple forms of communication can be best. For example, posters on information kiosks located in well-traveled areas can be an excellent way to update people on community happenings.

The choice of how to communicate will vary. During the emergency response period, police cruisers with loudspeakers may be the most effective means of getting information out. When basic utilities are beginning to be restored, radio and television make sense. Posters and billboards are useful, especially if local print media are not able to print and distribute papers. Face-to-face community meetings in neighborhoods are an important means of maintaining communication and a sense of connection for residents. In general, officials should use whatever means of communication that are available and accessible to the greatest number of community members. Face-to-face communication is widely acknowledged to be the most effective; electronic communication may be the least effective.

THE PROCESS OF RECOVERY

People, jobs, and housing are prerequisite to healing or rebuilding the social community. The key challenge in recovery is to bring these three elements back together in approximate proportion to how much they rely on one another, and then to help generate increases in each one until a critical mass is achieved that will build upon itself. Depending on the extent of damage to social, cultural, and economic elements of the community, the process may take years; however, there is no recovery without it.

What follows is an overview of activities that are inherent in this process— necessary, but not necessarily sufficient, for community recovery. Some steps are prerequisites to others; others have to move along in tandem, changing and adjusting as needed. It must be noted, however, that none of these activities has to be completed before work can proceed on the others. Recovery is not a "fix one sector completely and then move on to the next one" kind of activity. Each sector has to be addressed enough to permit other sectors to catch up, and then they can each be advanced a little more after that. Local administrators will lie awake nights struggling with the chicken-egg problem: "How can I get either in place without the other being in place first?"

Community Assessment

Essentially, the local leaders' challenge is to jump-start the recovery process so that private interests are sufficiently encouraged to invest their energy and resources into moving it along. Once private interests are confident that the community is on course to recover, they will be more likely to risk investing in it. In addition to stoking private interests' confidence in the community, local government must also provide basic services considered by many to be the foundation for successful communities (e.g., utilities, infrastructure, schools, health care, business districts, affordable housing, recreational facilities) or, if those services are contracted out, help private or nonprofit organizations provide them. Once private interests have made a commitment to the community, government must provide guidance as to what is acceptable and what is not, including where to build and what to build.

Consequently, it is necessary to do a careful, thoughtful, and accelerated assessment of where the community actually is and what future it likely faces. As stated earlier, local government will want to establish a reliable intelligence function. What would have been the most likely future for the community had there been no disaster? What will it take to get the community on a positive trajectory? What "critical mass" will be required? What is happening to similar communities across the country? Have any had the kind of success that your community would want to emulate? This is not easy work. It demands a clear-sightedness that many in the community would rather avoid. Thus, again, it is often useful to get outside help to see the community as it is, rather than as we remember it with selective recall or wish that it were. Unfortunately, consultants and state and federal officials are usually optimistic to a fault. No one wants to say that the best thing to do is probably to give up and move on. Nonetheless, it is possible to get people with a stake in the community to develop a fairly realistic appraisal of the community's future.

The simplest approach to this—and one with which many members of the community will already be familiar—is a SWOT analysis. What are the community's strengths, weaknesses, opportunities, and threats? Such an analysis cannot be performed in an hour or two by a large group over coffee and donuts down at the schoolhouse. Rather, it requires serious thinking by skilled people over whatever time is available. Too often, we have seen organizational and government officials conduct SWOT analyses and develop "strategic plans" in an afternoon with a group from the community or the organization. Doing so is "good for show" and may even provide some insights, but it rarely results in adequate analysis or realistic planning. Similarly, if local leadership has some strong ideas about what to do and simply pretends to engage in group planning, the more savvy people in the group will pick up on it in a heartbeat and any goodwill that might otherwise have developed will be turned into negative response.

Community Visioning

Following a careful assessment of where the community is at this time must come an even more daunting question: "What's possible for the future?" Although some communities tried valiantly to transform themselves into something different from what they were before the disaster, we did not see any that actually achieved the desired transformation, nor do we know what they might have done differently to effect the transformation. We concluded that it is extremely difficult to make the community into something that it was not, even in a decade of effort. Frankly, when the same people come together as have been together before, they have a tendency to think the same way, consider the same issues, use the same resources, and—not surprisingly—achieve the same results. This may be why some communities continue to hold out for the thriving "downtown" of yore. "Maybe if we tried harder, this time it would work." Doing the same thing over and over again and expecting different results is said to be the definition of insanity. If people know this (and it seems to be an oft-cited observation), then why do they do it? Perhaps because we seek affirmation of our own preferences and biases, and it is difficult for us to see how they color our thinking and decision making.

We do think that it is possible for communities to position themselves for transformation, but whether transformation can actually be achieved depends on many things coming together over a period of years. As World War II ended, Las Vegas was virtually nonexistent. Low taxes, massive private investments from a wide variety of sources, and a primary industry based on entertainment and lasciviousness—not to mention a strong desire on the part of several influential people (e.g., mobster Benjamin "Bugsy" Siegel, a member of the Meyer Lansky crime organization)—can do a lot to grow a city. On a more positive note, the Marshall Plan helped restore war-torn Europe into an economic powerhouse in a relatively short time.

The most promising approaches we have seen are those that build on what is already working in the community, augmented by new elements that have the potential to foster further development. A few other communities— St. Peter, Minnesota, and Biloxi, Mississippi, among others—have taken that approach with some success. St. Peter has positioned itself as a community that is complementary to nearby Mankato. It offers affordable and quality housing to individuals working in Mankato, where such housing is scarce. Biloxi was having some success with casinos before Katrina hit. Rather than seek some other form of industry afterwards, Biloxi chose to encourage the tourism industry by supporting additional casino development.

Goal Setting

At the same time it looks to see what *might* be done, it makes very good sense for a community to learn what *should* be done to ensure that recovery actions help it become more resilient, less exposed, and less vulnerable to subsequent events. All of this requires specifying some general goals or performance measures to strive for as it moves toward recovery.

Developing practical, achievable goals for the community is difficult. We have been to communities where some of the goals that were defined and the subsequent activities that were undertaken demonstrate the futility of shoveling sand against the tide. As stated previously, many mid-size communities have used valuable resources trying to recreate central business districts that resemble those of the first half of the twentieth century or those of very large cities—without much success. People interact with their immediate neighbors in neighborhoods that are located away from the city's downtown. They support shopping centers, grocery stores, and gas stations near where they live and work. The downtown is not a magnet for either social or work activities. Making it more attractive does not change people's living and working priorities.

Similarly, hundreds, perhaps thousands, of smaller communities have created "industrial parks" at the edge of town, put up a sign, and waited. Most are still waiting. Without a strong indication a priori that businesses want to locate in the community, it makes little sense to create a place for them to be. While it's true that businesses look for signals such a industrial parks before deciding to locate in a given community, they rarely locate in a community that doesn't already offer some level of opportunity to interact and "cluster" with associated industries.

Rather than pursue goals unrelated to its existing nature, a community must set goals that are congruent with the driving social and economic forces that are shaping it, as well as with its individual community's particular assets. This doesn't preclude attracting new industry or doing new things; instead, it argues that communities must assess and build upon their core competencies. Diversification has a stronger likelihood of being successful if it is related to the community; in which case it is probable that at least some of what is needed to support new enterprise is already present and, ideally, functioning well.

One extremely important goal for a community post-disaster is building hazard mitigation into its recovery plan. Those seeking to resurrect the community exactly the way it was before the disaster need to recognize that if the pre-disaster community was vulnerable to this disaster, chances are that, if it is rebuilt just as it was, it will be similarly vulnerable to a subsequent disaster. The challenge for local government and community members is to build *a bet-*

ter, more resilient version of what existed before the extreme event, while taking into account the community's essential nature and core competences. Books such as *Holistic Disaster Recovery: Ideas for Building Local Sustainability after a Natural Disaster* provide guidance on how to build resiliency into a community through hazard mitigation.[42]

It is also important to include all community constituencies in the goal-setting process. Community residents can sometimes have quite definite ideas about what they want and do not want. Listening to key constituencies (e.g., property owners, business owners, teachers, municipal employees) and discovering their problems, concerns, and needs is far more fruitful than sitting around a table in city hall either assuming or guessing what they want. Local government officials add insult to injury when they impose their own priorities on constituents whose perceptions and expectations have been ignored. At the same time, it is vital to move the process along quickly enough so that delays don't become dysfunctional. A leadership advisory team charged with moving the community forward must balance the benefits of participation with potential costs of the same.

Determining the Basic Strategy

Once the community assessment has been made and the goals defined, the leadership advisory team should spend a considerable amount of effort over a relatively short period trying to define a basic strategy for moving the community from where it is to where it wants to be. To do this, it has to conceptualize, outline, and evaluate the likely efficacy of various strategies, with the understanding that no community stops changing unless it is defunct. Defining such a strategy is, of course, considerably easier said than done. It means relying more on facts and on careful assessment than on hope and intuition. It means diversifying as much as possible: single-industry communities are vulnerable in ways that multi-industry communities are not. It means reviewing the strategy regularly and rigorously to evaluate the extent to which it is proving out. It means being bold enough to capture people's imagination, but not so bold as to appear out of touch with reality. Mostly, it means surrounding oneself with smart people who are not reluctant to say that the strategy is flawed if that is how they see it.

Whatever basic strategy is ultimately decided on, reality testing is important. Too often we have seen communities commit to a strategy that has only a slim chance of producing the desired outcomes. It is almost as though people think that believing in the efficacy of the strategy is enough to make it work. Of course, no one can ever tell in advance whether a strategy will work out as hoped, but we believe that it is extremely important to get objective, constructive criticism both before committing to a course of action and after. Success is more likely if the strategy is revisited often enough to ascertain whether it is producing the hoped-for results or needs to be reworked.

Thus, the basic strategy, which is necessary to help ensure that the community can effectively use the resources it has to move toward recovery, is really a general course of action, one that can be adapted as situations and needs change. Either the leadership advisory team or another team must be responsible for appropriately sequencing and staging activities to meet current and emerging needs. It should identify critical opportunities for "seeding" the community with projects and providing assistance to private groups to guide subsequent private actions.

As individuals, organizations, and government take action, contingent events will play out, conditions will change, and some uncertainties will be resolved as others emerge in their place. No matter how carefully or deliberately we have planned, unexpected or undesirable outcomes or side effects may arise from some action that has been taken. New obstacles or new avenues to goal attainment will appear, like unwanted epiphanies, demanding a change in strategy to accommodate them. For example, until the factory opens, it is impossible to know whether a local employer will be able to make good on his or her promise to rebuild and reopen. Until former residents begin to return in numbers, it is not prudent to count on that happening. Until a firm commits to building in the community's new industrial park, it is unwise to assume that it will.

To be effective, then, planning strategies and programming activities have to be a continuing process, particularly in times of great uncertainty. Plans must be flexible; they must provide for branching out into different directions at various junctures if such revision is called for. Adaptation does not mean abandoning the plan and starting again from scratch; it requires working with the reality of the situation to increase the probability of reaching the intended goal. And to this end it is critically important that some group in the community be responsible for continually monitoring the recovery process to identify unexpected events, outcomes, and trends, and to assess how well the chosen strategy is working toward reaching the community's recovery goals.

As stated previously, we urge decision makers not to make early decisions that commit them irrevocably to one or another course of action. "Soft paths" are preferred to "hard choices." Making hard choices early limits one's options later, just when they are needed. Some people will press continually for an early decision because ambiguity is intolerable to them. A former employee of one of the authors once pleaded, "I don't care if it is right or wrong, but I really want you to make a decision about this so I can stop thinking about it." The author decided to replace the employee; ambiguity is part of life.

Post-Disaster Community Planning

We have not said much here about traditional land use and facility planning. Traditional approaches to community planning are inadequate for planning post-disaster community recovery. They are, however, important as an expression of those community goals that can be displayed on maps: for example, land use allocation, distribution of facilities and physical infrastructure. These approaches are very much like a blueprint for a house. They show what it should look like, but do not say much, if anything, about how to get from a vacant lot to the final house with people living in it. They will help guide government decisions about where to put facilities and infrastructure, and will influence private decisions about where and where not to build.

The planning we suggest local governments embark on embraces land use and facility planning, but it is much more. It addresses a broader array of goals, and is oriented toward devising a course of action that is intended to get us from where we are to where we want to be. It consciously addresses the existence of contingencies and unknowns, and accommodates the need to change and adjust in anticipation of events and in response to changes in circumstance.

CHAPTER 19

PITFALLS IN PLANNING FOR COMMUNITY RECOVERY

"Life is under no obligation to give us what we expect."

Margaret Mitchell

The battle over whether to demolish public housing in New Orleans post-Katrina reflects pre-hurricane discontent with the housing and associated ills such as crime, beliefs that such housing is needed by the city's poorer residents given a reduced housing market and escalating prices, and perceptions that the city will be "better off" with public housing gone. Photograph taken in New Orleans in October 2005.

*P*ITFALLS EXIST IN EVERY ENDEAVOR. It is particularly easy to fall into them when under pressure to address a spate of new problems and urgent needs. Here, we describe the primary pitfalls that we became aware of while studying post-disaster communities.

CONFUSING MUNICIPAL RECOVERY WITH COMMUNITY RECOVERY

It is important to distinguish between municipal and community recovery. There is a significant difference between the legal entity of the local government corporation, with its defined spatial boundaries, and the "natural" community or communities that it serves, and the two might not coincide. The complexities of American urban settlements and the arbitrariness of municipal boundaries within those settlements result in situations in which municipalities and communities are seldom coterminous. Several communities may exist within one large municipal jurisdiction; similarly, only part of one or more communities may exist within a municipality.

Like any other "business" enterprise, local governments often suffer great losses from extreme events. In Princeville, North Carolina, and in Gulfport, Mississippi, the city halls were ruined and made uninhabitable by extreme events. In Homestead, Florida, the city-owned electric utility was completely destroyed. Across the country, municipalities suffer significant declines in revenue and revenue-generating capacity as extreme events destroy the real property tax base and reduce sales tax revenues. However, repairing and restoring municipal buildings, infrastructure, and enterprises, while necessary to community recovery, does not necessarily guarantee recovery of the community, of the neighborhoods within it, of families, and of individuals.

Some officials fail to distinguish clearly between recovery of the municipal enterprise and recovery of the community, often to the detriment of the latter. We think that these officials are probably using a definition of recovery that is limited to replacing the built environment and municipal enterprises and excludes reestablishing interrelationships among the members of the community and between the community and the local government. Perhaps they think that community systems will restore themselves to pre-event order if the city simply provides infrastructure and works to rebuild its revenue base. Or perhaps they believe they have no mandate or responsibility for doing more. More often than we can count, local officials said to us, "Don't tell anybody I said this, but this [insert kind of extreme event] is the best thing that ever happened here." We were never sure of exactly what they meant, but we think it's something like this: "Nobody died, but we were able to get rid of lots of old housing; we have all new [schools, streets, wastewater treatment facility, water supply system, administrative buildings, parks—check all that apply]; and they were paid for by the federal government. The city is essentially debt free. How much better can it get?"

The local government managers we talked with generally see municipal and community recovery as two sides of the same coin: the community needs an effective local government and the local government needs a community with a relatively strong economy to provide jobs, a secure future for residents, and a tax base sufficient to meet the revenue needs of the local government. We think most local officials have two implicit goals: recovery of the municipal enterprise and helping the community system to recover. They are committed to the latter goal to the extent they understand what constitutes recovery, have the resources to pursue community recovery, and believe that they have the responsibility and the mandate to do so.

ASSUMING EVERYONE HAS THE SAME RECOVERY GOALS

People in different roles often have different perspectives and do not necessarily share the same goals. The city government might want to have a bigger, better retailer selling such and such. But the smaller retailer who had been selling such and such wants to be back in business making money; he doesn't want someone else providing that service. City officials may believe it makes good sense to remove housing from a particular area that is flood-prone and turn that area into a nature preserve, although the people who live there might want to stay.

The continuing conflicts in New Orleans following Katrina drive home this point. Rebuilding the Lower Ninth Ward may make no sense from the perspective of creating safe housing in a safe environment. After all, the area flooded and was inaccessible for months after Hurricane Katrina. Now it is wiped almost clean of homes with only remnants remaining—slabs, porches, and parts of fences. Some people, however, see the rebuilding issue as one of individual property owner rights. Nearly 60 percent of the homes in the Lower Ninth Ward were owner-occupied before Katrina, compared to 46 percent of homes in New Orleans overall.[43] From their perspective, the area would be safer if the Army Corps of Engineers were to build and maintain a levee system consistent with a category 5 hurricane. After all, they say, other flood-prone areas of the world (e.g., the Netherlands) have much more resilient flood protection, and no one questions individuals' rights to live in these otherwise unsafe areas. The conflict over rebuilding the Lower Ninth Ward continues three years after Katrina, with no reasonable solution in sight.

Conflict sometimes emerges when mitigation projects are undertaken following the disaster. In Grand Forks, an expanded and enlarged levee system was planned for construction. During the design process, a dispute arose between those who were scheduled to be inside the flood protection levees and those who were going to be left outside. Those who were to be on the inside of the levee argued (along with local officials) that it would be too expensive and impractical to put everyone inside. Those proposed to be left outside the levee argued that as long as they were helping to pay for it, they had as much right to be on the inside as anyone else. Everyone believed they had a valid point. Such conflicts can change who is elected to what positions in a community. More than one mayor has been replaced because of something said or done in the aftermath of an extreme event.

People from the state or federal government sometimes have objectives that are slightly different from those of local officials with respect to the disaster. Local officials want to see their community recover. Individual staff members from other levels of government typically share that concern but are charged with ensuring that regulations are followed, monies are spent and accounted for appropriately, and projects are completed. Their primary job is to administer those programs efficiently and effectively. While we might hope that the goals of all levels of government would dovetail, that kind of congruence is not assured.

At the same time, local officials sometimes seem less concerned with the plight of individuals and individual businesses than with aggregate measures of population, employment, income, and the removal of physical manifestations of the disaster. Most individuals who had losses were concerned, at some level, with the well-being of the community, but their main focus was usually on themselves, their families, and their friends and neighbors.

Clearly, the criteria for evaluating recovery vary across levels of analysis and perspectives. It is difficult, often impossible, to get everyone to agree on a specific course of action, since it is difficult to create solutions in which everyone gains or at least breaks even.

LOSING TRACK OF THE BIG PICTURE

It is particularly difficult to local officials to see the big picture—to keep an open mind on issues, to identify and explore options, and to make objective decisions—when their city has been devastated and the struggle for survival and recovery is being waged. Several things work against them when they try.

First, people will want and expect the community to return to what it used to be—what they consider to be "normal." Most people really dislike ambiguity, much less being torn irrevocably from their accustomed surroundings and routines, and many work diligently to restore what had been so that they can pick up where they left off. But unless there were few social, economic, and political consequences from the event, that cannot be done. Communities change every day; if they do not change, they do not survive. Usually the change is slow enough for people to acclimate to it without a conscious recognition of their adaptive behaviors. In times of upheaval, however, the changes will be rapid and unsettling, and so local officials must be able to communicate clearly and convincingly that life will go on and that everyone is working hard to ensure that the changes will be for the better.

Second, it is extremely difficult to step back to gain a larger perspective when there are urgent demands requiring immediate attention. Consequently, local officials need regular reality checks during the aftermath of an event, and it is best to get them from people who do not have a stake in local outcomes. Consultants with integrity—those who are more interested in solving problems than in selling solutions—are useful, particularly when the "contract" calls for the consultant to tell the truth as best as he or she sees it. One of the most dysfunctional things we saw in communities trying to recover was a phenomenon we call "group shift," in which people were induced to believe in something entirely implausible, and to engage in and promote highly risky personal investments. Also widespread and problematic was groupthink, in which people silence their disagreements to present an illusion of unanimity. Unless conflicts are openly raised and addressed, they will fester until they ultimately threaten to undermine recovery efforts.

Finally, once people match up a solution with a problem, they tend to assume that things will work out exactly as they imagined, hoped, or visualized. Having made the causal leap that "if I do x, then y will for sure happen," they charge ahead with vigor and determination, oblivious to flaws in their logic, possible pitfalls, unknown probability distributions, unanticipated consequences, and changing circumstances that may make the plan impractical. Even worse, they often become resistant to well-meaning criticism. Often they fail more because they find a solution to the wrong problem than because they fail to solve the correct problem. Sometimes decisions have to be made quickly, but not as often as is generally thought.

In general, then, we advocate the use of dispassionate analysis. The most successful communities took the time to think through their plans and proposals, and to subject them to serious scrutiny from multiple perspectives by people committed to sound planning and decision making. Again, the value of an external review cannot be overstated.

NOT HAVING A STRATEGY FOR GETTING FROM HERE TO THERE

Too often, community planning focuses on what the community can or should become without developing a basic strategy for getting there. A plan that doesn't include action steps, performance measures, and criteria for decision making is little more than a wish. Both the strategy and the goals for the community must be sufficiently flexible to accommodate the vast number of unanticipated contingent events that are likely to arise between now and then. No one can control all the variables, and local officials do not call all the shots about what happens in the community. Communities are, after all, self-organizing systems, and municipal policies are only one factor, albeit an important one, in what happens and when and where it happens.

That said, it also makes sense to have requirements to guide public and private investments so they are supportive of the community's longer-term goals. Without criteria against which to evaluate alternative actions, officials can do little more than react opportunistically on a case-by-case basis to what is or might be available, simply in the hope that whatever brings resources into the community will lead to good things.

The challenge of developing a proactive strategy for physical, social, and economic recovery is that local officials are expected to hit the ground running after an extreme event. Money to assist the recovery effort is based on the community's ability to provide project plans. If the project plans are to be based on an integrated strategy, the local government has to have a strategy roughed out in advance of the disaster or has to develop one quickly after the event—even as it is focused on meeting very immediate and very urgent needs. In today's jargon, it demands effective multitasking.

NOT ALLOWING FOR UNANTICIPATED OR UNCONTROLLABLE EVENTS

We are constantly amazed at how many people believe that, once they have set a course of action, things will work out just as they had planned or imagined. Such a fantasy is probably comforting—at least until it falls apart. But in reality, we do not live in a world of certainty. We live in a world of uncertainty and, often, ignorance. Even our best-laid plans are subject to the impacts of unanticipated events and events beyond our control.

No matter how diligently you rake in the autumn, you can't get every leaf. Likewise, there is no way to control every outcome in the recovery process. Many things that happen during the aftermath have probabilistic outcomes. Accurately predicting the outcome of a series of sequential events, each of which has a probabilistic outcome, is, at best, unlikely

Because the consequences continue to unfold for months, sometimes years, after the event, it is useful to try to identify what is happening in the community at any given time. What trends appear to be emerging? How likely is each to continue? What apparent anomalies are manifesting themselves? Are they the precursors of trends to which one should be attentive?

Many actions and consequent outcomes are in the hands of others. Decisions are made by state and federal officials, corporate executives in other states, individuals within your own community, and many other people acting generally independent of local government leadership in the community. Local government can influence outcomes, but it cannot control them. As we have said, local communities tend to behave in the manner of self-organizing systems. They are partly the consequence of local government policy and action, but they are largely the consequence of individual and organizational choices.

We think it is possible for state and local government to seed the community with projects and investments that help to swing decisions by others toward the desire path of community recovery. We think, too, that local governments can use "pricing" to encourage some things and discourage others. For example, if a state government does not want development pressures in a pristine wilderness area, one way to increase the cost of developing the land is to not put roads or freeway interchanges in the area. Similarly, local governments can make it easier to do what is desirable and harder to do what is not desirable. It is a way to use the power of the public purse and the police power of local government—that is, the authority and duty to do what is best to protect public health, safety, and general welfare—to leverage desired paths and outcomes.

ASSUMING IT WILL WORK HERE BECAUSE IT WORKED THERE

It is tempting to learn what other communities did to produce good outcomes and to try to replicate those actions in the hope of achieving the same outcomes. But in looking at communities across the country that have experienced disasters, we have come to see that context is critical in determining whether a solution that worked in one community will work in another. In short, just because it worked or didn't work there doesn't mean that it will work or not work here.

The lesson is that there is no cookie-cutter solution to community recovery. As we have noted several times, some things are essential but not necessarily sufficient. It is critical to get the basic utilities, infrastructure, and services up and running. It is critical to help ensure adequate housing and to facilitate private efforts to get needed private services back in operation. Beyond that, however, recovery efforts should be determined according to the unique context of the community and what someone did somewhere else.

FORGETTING ABOUT PATH DEPENDENCY

In more than one city across the country, the streets in the old part were laid out by deer trails, hardened as they became cow paths, and finally institutionalized with pavement. The original typewriter keyboard was arranged as it is so that the letter arms would not get stuck on one another as they struck the ribbon to make an impression. Only a few die-hards still use mechanical keyboards, but the arrangement of computer keyboards remains the same. Along these same lines, what local government does early in the aftermath will set a pattern for what follows. "Soft paths" is a term sometimes used to describe courses of action that are revocable and from which retreat is possible without devastating consequences. Soft paths are desirable when the decision maker sees too much uncertainty to commit irrevocably to a specified course of action. The challenge is to set a reasonable course of action that leaves room for more sensible options to be explored; these options might then be implemented as more information is obtained and uncertainty is reduced to manageable levels. Once the concrete hardens, it is a little late to change direction.

This ties to the notion of not being rushed into decisions. Pressure will come from many sources: individual citizens, business owners, school districts, and other government agencies, to name a few. People will want to know what is being done NOW to help them. They will see the aftermath of an extreme event as an opportunity to secure resources they have long desired, and to accomplish goals that—up to now, for them—have been unattainable. They will offer solutions to problems that don't exist, to problems that are urgent but not necessarily important, and so on. Their efforts to persuade will be genuine and rooted in what they think is best. The task before the local government official is to sift through the various problems that have been identified, the various solutions that have proposed, and the myriad offers of assistance that have been received to create a series of doable plans and carefully evaluated decisions.

Deliberate decision making takes time. Officials cannot allow themselves to be rushed by statements that "recovery requires fast decision making." At the same time, they must be aware that the time needed to make the most comprehensively considered decision does not exist. Thus, they must make decisions faster than they may have ever made them before, while taking care to minimize commitments to paths that may be irrevocable and costly. Engaging the services of individuals who do not have a vested interest in the outcomes and who are skilled in decision making may increase the probability of making good decisions. So, too, may simply being aware of the need to make relatively fast decisions that align with "soft paths."

PITFALLS NOTWITHSTANDING ...

It seems to us that community recovery is a lot like sailing to a specified destination while unknown, unmapped reefs lie dangerously hidden underwater and the wind changes in velocity and direction every few minutes. Everything requires careful orchestration, which can be accomplished only by bringing the key elements of the community together. Local government cannot do it alone, but it should be the conductor. Some things have to happen simultaneously or, at least, in tandem (e.g., jobs and housing), while others can happen sequentially (e.g., schools following jobs and housing). It is necessary to maintain continual monitoring and frequent feedback of the community system to identify disjoints and gaps that require someone's immediate attention. Community recovery will be herky-jerky with too much of some things and not enough of others, with the entire thing in continual flux, and with a never-ending stream of urgent problems crying out for attention.

THE PRINCIPLE OF INCOMPLETENESS

For as long as they exist, communities are constantly changing. Neither the community nor planning for its development is ever completed. It is neither necessary nor possible to design the ultimate community, nor can we reliably imagine life half a century from now. We think that the challenge is to address the problems we can identify while not precluding continuing adaptation to subsequent changes, problems, and opportunities. As is the case with individuals, communities do well that develop the capacity to adapt to changing conditions in addition to sustaining their core competences.

CHAPTER 20

BEFORE THE NEXT DISASTER: HAZARD MITIGATION

"To be prepared is half the victory."

Miguel de Cervantes

WE URGE LOCAL OFFICIALS TO view the community as a system rather than simply as a collection of facilities, structures, and people. Their objective should be to build resiliency into the system so that it can absorb or reflect shocks and continue to function. In the words of the old commercial, the community should be able "to take a licking and keep on ticking." Increased resiliency can be attained in several ways, yet they all come down to reducing exposure and/or vulnerability. Systems theorists tell us that systems survive only when they have an array of coping mechanisms at least equal to the array of challenges thrust upon them.[44]

For some years now, FEMA and others have been assembling evidence to demonstrate that hazard mitigation is worthwhile. Mitigation reduces injuries and deaths. In the case of public infrastructure, mitigation saves taxpayers money by reducing losses. It also often saves building owners money and enables them to get back into business quickly. However, we think that the most important and valuable effect of hazard mitigation can be summed up as follows: *The greater the initial losses from an extreme event in terms of injuries, deaths, and damage to the built environment, the more vulnerable the community is to cascading adverse consequences in its social, economic, and political fabric.* Thus, mitigation that reduces losses to life and damage to the built environment reduces the extent of cascading adverse consequences, thus making community recovery more likely and more likely to proceed quickly.

Put differently, communities that devote resources to mitigation are more likely to reduce the immediate and immediately following consequences of the event. As a result, systemic community consequences, ripple effects, and ripple reverberation effects are less likely to be severe. When the built environment suffers less damage and when fewer people die, fewer families move away and fewer businesses go out of business; there is also less lingering terror, less social disruption, and generally fewer adverse consequences for the community. The fewer and less debilitating the consequences, the more likely it is that the community will recover in relatively short order. Hazard mitigation should be built into the routine processes of every community vulnerable to an extreme event.

We do not believe that one should ever spend more to solve a potential problem than it will cost if that problem comes to pass. Nor do we believe that elaborate cost-benefit analyses are always necessary when making mitigation decisions. Most often we approach what we call "the hazard mitigation investment decision" as, essentially, a risk management problem. We ask ourselves, "How much are we willing to spend now to keep the worst consequences of some uncertain future event from happening?" That implies a cost-benefit analysis, but it does not demand that we assume the event will occur. In that respect, mitigation is a lot like buying auto insurance. We don't expect a serious auto accident, and we don't know whether we will have one, but we can judge how much we are willing to spend to keep from suffering the extreme financial consequences associated with having one.

Communities can mitigate their risk by reducing exposure, increasing robustness, building redundancy into their facilities and practices, increasing diversity, and preparing for rapid replacement. In business, this approach is often called developing and implementing loss-prevention strategies. Hazard mitigation can be expensive, and some think it out of reach for their communities. However, some precautions that reduce the likelihood of extensive losses are not necessarily expensive. In fact, some precautions may even generate revenue.

CONDUCT A MACRO RISK ASSESSMENT

The first step in reducing the likely consequences of an extreme event is to conduct a macro risk assessment, defining the worst things that could happen to the community—the things that would cause no end of anguish and difficulty. When doing this, it is less useful to think of causes (e.g., earthquake, flood, and tornado) than it is to think in terms of consequences. Various extreme events can yield the same or similar consequences. Recently, the staff of a large sewerage district was working on its first macro risk assessment. One staff member suggested that one of the very worst things that could happen would be a train derailment that spilled the gaseous chlorine used by the district. Others added that it would not take just a derailment to cause a large chlorine leak: there were other possibilities. They took their concern to management, and the district decided to eliminate the possibility of a chlorine spill: it spent a million dollars in capital outlay to replace the process that used gaseous chlorine with a process that uses hypochlorite, a safe alternative with lower operating costs.

Focusing on only one potential cause of a gaseous chlorine leak may have led the management team to dismiss the cause ("That's so unlikely as to be ridiculous") and therefore not act to mitigate against the possible consequences of a derailment. Multiperil thinking, however, is more likely to generate robust analyses of extreme events and their likely consequences. Besides thinking about how different perils might yield the same or similar consequences, officials should also consider the results of multiple, sequential extreme events: hurricanes followed by floods, earthquakes followed by fire. As we heard from many in New Orleans, they were ready for Hurricane Katrina. They just weren't ready for the catastrophic flooding that followed.

Of course, it may be helpful to think in terms of the likely consequences stemming from a specific extreme event. "What would happen if the levee fails or the river overtops it?" It is probably better to think about levees as being hazards rather than the things that keep the community safe. Almost all the serious flooding we have seen in the communities we studied resulted from the failure or overtopping of a levee or flood wall. It makes sense for officials to look at what happened in other communities that experienced events to which their communities are vulnerable. "What would happen if we had an extended power outage?" "Where are the realistic sources of danger in our community?" And they should not overlook mindless acts of destruction: most communities are unlikely to be the target of foreign terrorists, but, as in the case of the bombing of the Alfred P. Murrah Federal Building in Oklahoma City, domestic zealots and individuals with severe psychological problems can cause serious damage.

As part of the macro analysis, it makes sense to identify events that would cause severe problems, cost a great deal to repair, take a long time to repair, or be impossible to repair. These events should be ranked in terms of the severity of their consequences rather than the likelihood of their occurrence. Although most consulting risk analysts will encourage officials to think of the risks that are most likely to occur to them or their organizations, we think that it also makes sense to look at both the most likely *and* the most ruinous. This does not mean that the staff should begin worrying about how to protect the community against collision with an asteroid greater than one kilometer in diameter: we don't yet know how to do that, and even if we did, mitigation would presumably take place primarily at the federal level. It does mean, however, that the staff should focus on the rare EF 4 or 5 tornado; the category 3, 4, or 5 hurricane; the magnitude 6 or 7 earthquake; a massive explosion in the city; a wildfire raging through urban-wildland interface; or the Humongous Auto Company closing its assembly plant in the community. These are truly rare events for any community, but each has happened in one or more communities in the United States during the past decade. And identifying the extreme events that could strike a community should not take more than an hour over coffee.

WORK TOWARD COMMUNITY RESILIENCY

Local government can use its annual budget to increase community resilience. Mitigation should be one criterion in every capital improvement spending decision.

The Built Environment

When physical improvements are to be made in community infrastructure, their resistance to extreme events should be one criterion for selecting location, design, materials, and interconnectedness with other infrastructure elements.

Unfortunately, communities today consist entirely of existing buildings; new buildings are either on the drawing board or under construction. It is not always easy to induce private property owners with existing buildings to take precautions against hazards. Requiring retrofit usually generates significant resistance, sometimes becomes political, and can result in the watering down of requirements.[45] For private property owners to retrofit properties to make them more robust, the investment has to fit into their business strategy and pocketbook.[46] Thus, it is far easier to require that precautions be taken for new construction, and to make compliance with up-to-date codes and regulations a condition of construction. Owners who know costs in advance of design and construction can more easily incorporate those costs into their investment decision. But since improving codes to ensure greater resiliency will apply only to new construction, it will take a long time for existing buildings to be made disaster-resistant.

Many communities across the nation require that buildings be brought up to contemporary standards when improvements or repairs exceed some threshold. It makes sense to require compliance with hazard mitigation practices at the same time that compliance is required with changes in specialty codes, such as fire resistance, plumbing, and electrical. Figuratively speaking, local officials have to think about whether they should be planting seeds, bending twigs, or moving trees. We suggest that local governments identify points of leverage they have on privately owned facilities to encourage owners to build mitigation strategies into plans for subdivisions, new buildings, and the other elements of the built environment. Local governments can increase the costs of doing things that are not in the public interest (e.g., building in a floodplain) and decrease the costs of doing the right things (e.g., building on higher ground). In addition to monetary costs, these can include regulations and requirements, and the amount of red tape that must be dealt with in order to do less desirable things.

Social and Economic Resiliency

We know more about making buildings and infrastructure resistant to extreme events than we do about making communities more resistant and resilient. Obviously, the likelihood of social and economic disruption in a community is significantly reduced if the built environment sustains relatively little damage from an extreme event. However, the community is also likely to sustain less social and economic damage if it was "healthy" before the disaster.

Thus, it is important for local officials to strengthen all elements of a community before an extreme event occurs. This means, in part, ensuring a diversified economy with good prospects for the future.

It also means reducing or eliminating dysfunctional components and relationships. It is always daunting to make headway against crime, poverty, racism, and other social maladies, but we have seen disasters open doors for cooperation and collaboration among disparate groups in many smaller communities. And even though the spirit of cooperation and collaboration declined in the following months, some elements remained in most places. This suggests the possibility of strengthening communities around common concerns—provided, of course, that someone takes the leadership in doing so.

We have also seen some communities where disasters exacerbated simmering conflicts between haves and have-nots and between races; sometimes local government actions inadvertently contributed to those conflicts. In one community, low-income African-American families, displaced from their homes by flooding, were bused from shelter to shelter to make room for mostly white, mostly better-off residents of a senior residential care center. It may have made sense logistically to transport the elderly just a short distance to a specific shelter and to transport the families who were younger and in better health to a different shelter, but it was not perceived that way by the people who were being shuffled between shelters. The hostility generated by the decision lasted for years. If divisions already exist, the local government should not exacerbate them by overlooking their importance in an emergency.

Local officials need to remember that some communities begin their recovery at the "starting line" while others begin ahead of the line or behind it. Our research suggests that there is little hope for those communities that start very far behind the line.

Reduce Exposure if it Makes Sense

Reducing exposure can be relatively inexpensive. In New Orleans, dozens of backup electrical generators had been "conveniently" installed in the basements of commercial buildings. They flooded and were useless. Similarly, floodplains should be used for parks, soccer fields, and open space, not for schools or other public facilities. When a facility becomes outmoded and needs to be replaced, its replacement should be put it where it is less exposed to the kinds of hazards to which the community is prone.

Sometimes a window of opportunity opens, making it possible to make major headway in risk reduction in a short time. But it is neither possible nor necessary to reduce exposure overnight. Mitigation should be an everyday activity, not one that is limited to the months following a disaster. To this end, priorities should be established.

Reduce Vulnerability of the Built Environment

If the desire for a beachfront home exceeds the fear of potential hurricane damages to it, the owner can "harden" the structure or insure against the likely loss. Hardening may mean making a building either more rigid or more flexible. Since Katrina, hospital owners along the Gulf Coast have been hardening their facilities, often in relatively simple ways. For example, they are replacing standard window glass with "184 mph" glass to reduce windows blowing in and sending glass shards through patient and administrative rooms. They are also removing roofs with pea gravel toppings applied over rubber membranes or composite roofing diaphragms because hurricane-force winds pick up the gravel and turn it into shrapnel that takes out windows in adjoining buildings and damages automobiles in adjoining parking lots.

Structural engineers continue to design buildings to resist high winds, storm surges, and earthquakes. Even if the building continues to stand, however, nonstructural elements (e.g., HVAC, plumbing, electrical) must also be made more robust so that people are not injured by falling ceiling tiles, light fixtures, and bookcases, or by explosions from leaking gas pipes. Although nonstructural elements and building contents are often more expensive to replace than structural ones, it is *less* expensive to make them less vulnerable. A number of relatively simple fixes are available to building owners who wisely choose to protect their buildings' nonstructural elements and contents against shaking, wind, water, and other hazards. For example, rather than design water and sewer systems using "wheel and spoke" templates, they can be designed as distributed, interconnected networks, which reduces the vulnerability of the system: as it is less likely that all the sites will be damaged, it is less likely for the system to become totally inoperable. Contents can be anchored or relocated to less vulnerable areas of the building.

Build in Redundancy

Sometimes redundancy is useful and prudent. Redundancy is putting at least two entrances in a building to increase both convenience and accessibility in case one is blocked. Redundancy is having a working backup generator and sufficient fuel to run it for more than the recommended seventy-two hours. Redundancy is drilling a well or having a portable water source in case the city's water is unavailable and the system requires potable water. Redundancy is backing up data regularly and frequently and storing it off-site, ideally at more than one location. Redundancy is cross-training employees to handle multiple tasks, especially in the face of an extreme event. Redundancy is involving more than one set of staff members in emergency drills, increasing the likelihood that knowledgeable people will be available when needed. Sometimes, redundancy simply means having alternative strategies available if one proves untenable.

Spread Your Risks

Retail businesses with multiple locations have a much better chance of surviving an extreme event than do businesses with one location.[47] Organizations with several warehouse facilities are at less risk than those with everything in one place. Communities that have several commercial nodes have a better chance of having some of them survive an extreme event than a city with a single commercial center. Communities with diversified economies are likely to fare better when one industry faces a long-term downturn or has its facilities badly damaged by an extreme event. Communities should think in terms of supporting a portfolio of businesses in their communities, some of which complement and support one another, and some of which are wholly distinct.

Prepare for Rapid Replacement

One way to address vulnerability is to accept that some facility or part of a facility need not be hardened to withstand every event, but to make arrangements for it to be replaced or repaired quickly. Most organizations that use large computers have arrangements with one or more vendors to have the computers up and running within some specified short time after they crash. Most organizations also have an inventory of equipment or parts that are likely to fail or that are absolutely critical. For many business and production organizations, this is standard operating procedure: "We can't keep it from happening, but we can get it operational again in no time." Of course, maintaining duplicates of everything, while consistent with the strategy of redundancy, may not be cost-effective. "Stuff" needs to be warehoused somewhere. Warehouses and the land upon which they sit cost money. Where this expense poses a problem, communities may want to establish agreements with outside vendors and other communities to acquire needed resources in short order.

It is important to standardize equipment, such as pumps, so that they can be replaced quickly and easily. If a special type or brand of equipment is being used in a critical facility or capacity and is no longer obtainable or even being made, adequate backup must be available nearby because replacing that specific item will be impossible.

Some hospital officials we interviewed in areas prone to hurricanes and tropical storms reported that they have standing contracts with firms to supply them with fuel, food, and other supplies in the event of a storm. When securing these contracts, however, it is advisable to ensure that the firm is not overcommitted to provide the same service to other jurisdictions likely to be damaged by the same event.

EMPLOY MANAGERIAL MITIGATION

Not all mitigation has to do with physical facilities. We have coined the phrase "managerial mitigation" to describe the application of management practices to reduce the likelihood of direct losses from extreme events. Here we suggest several steps that we think should be taken as soon as possible if they have not been taken already.

Insure the Assets

Traditional insurance is extremely important for local governments and quasi-governmental agencies. Local governments sometimes "self insure," which is to say that they don't insure at all. To be eligible for federal disaster assistance, local governments must comply with various requirements for insurance. The rules change from time to time without much fanfare, so it is incumbent on local government financial officers and risk managers to be attentive to the rules and to changes in them.

The chief financial officer (CFO) of one Florida city had been through a devastating hurricane in another city before she changed jobs. The former city wasn't insured against hurricanes, but she made sure that the new city one was. To obtain the desired coverage, several neighboring local governments worked together in the insurance market. Thus, when a hurricane struck her new city and did considerable damage, the city did not suffer major financial trauma. It had both good insurance and a CFO with hurricane experience who knew what to do. The insurance enabled the city to begin repairs and replacement almost immediately.

Insurance is particularly important for private organizations. Governments and nonprofits can receive federal aid, but private organizations and businesses are not eligible for disaster assistance. Small businesses that do not qualify for commercial loans to reestablish their businesses may qualify for loans from the Small Business Administration (SBA); however, as we have said earlier, taking on such a loan is not always a good idea for a small-business owner because it increases that owner's debt load. If the business's market has changed significantly because of the extreme event, there may be inadequate trade after the event to cover the business's cash flow needs.

Hire for Extreme Event Experience and Prepare the Staff

While the example of the Florida CFO speaks to the advantages of being insured and having a competent CFO, it also speaks to the value of hiring people with extreme event experience. In many respects, there really is no good substitute for experience. Individuals who have been through an extreme event know what it takes to survive. They know the challenges that will be faced, and they have some ideas about how to face them. Accordingly, they find it easier to plan, anticipating both the occurrence and the likely consequences of an extreme event. Rather than getting caught up in the probability of the event occurring, they approach it by reasoning, "If it happens, bad things will follow, unless we're ready. Where's the harm in being ready?"

Chances are that a local government staff has been trained for emergencies, but it should also be trained for dealing with the aftermath. Problems invariably will cut across departmental lines and, if they are to be addressed effectively, require collaborative, coordinated problem solving and service delivery. It works much better to get that experience and training before the community is collapsing around city hall. Staff members should be comfortable using integrated communication networks; any and all agencies likely to interact in the wake of an extreme event and afterward should have and know how to use communication technology that is interoperable.

Put Hold Harmless Provisions in Place

It is important for local governments to work together to have their state enact hold harmless provisions for communities that suffer disasters. It may be easier to do so after a community finds itself devastated, but it makes more sense to do so beforehand.

As suggested earlier, one such provision is for the state to use the previous year's school attendance when calculating state aids to education in school districts where enrollment has decreased dramatically. If the decline in attendance is likely to be permanent, then reductions in school aid might be gradual over several years to facilitate the local district's efforts to adapt.

A second such provision would hold local jurisdictions harmless for losses to sales tax revenues or other centrally collected, locally shared revenues following a disaster. Such an arrangement was made for Homestead, Florida, after Hurricane Andrew.

Develop Mutual Assistance Agreements and Have Contracts in Place with Reliable Vendors

Most local governments have mutual assistance agreements with other communities for public safety activities. Such agreements might be expanded to cover post-disaster computer use, warehousing, and the provision of temporary supplies of gasoline and other consumables.

Communicate the Risks to the Public and Train the Public on What to Do

While one of the authors was teaching at military bases in Hawaii, an Alaskan earthquake triggered a tsunami that headed for Hawaii. Broadcast media directed people to turn to a specific page in the telephone book, where specific, easily understood instructions could be found on where to go and how to get there in the event of a tsunami. The city of Honolulu was well-prepared. The bus system immediately began to transport people to higher ground, and the broadcast media provided everyone with up–to-date information. Fortunately, when the tsunami struck, it was only four inches high, much to everyone's relief and to the great amusement of people who print current sayings on t-shirts. The only untoward event occurred when one person was directed to a beach swarming with Japanese tourists: as he ran toward the beach with a bullhorn in hand, he was heard to shout back rather urgently, "What's the Japanese word for tsunami?"

PLAN NOW FOR WHAT TO DO IN THE IMMEDIATE AFTERMATH OF THE DISASTER

It is becoming popular to devise pre-disaster plans for recovery. Unfortunately, it is impossible to know how a disaster will unfold. No one wants to spend time devising a plan that falls apart the first time some unanticipated contingency arises. How can a local government plan sensibly for recovery before the event? The answer is relatively simple: officials need to plan for the things that they know have a high probability of happening.

Process Planning

Whatever the nature of the extreme event, whether it is an airplane crash on a sports venue, a toxic gas cloud, or a flood, there has to be a process in place for getting the relevant actors together and ensuring that they are all rowing in the right direction. There is no time for sorting out jurisdictional squabbles between departments in the same government or even between adjoining governments and governments at different levels when the castle is on fire and the barbarians are at the door.

Depending on the nature of the event, many activities will have to be undertaken. Thus, the local government's process plan should specify the roles, responsibilities, and degree of authority of everyone who is to be involved should an extreme event occur. Since personnel tend to change frequently, the plan should be updated regularly; those who have been assigned specific roles should be trained and retrained; and all relevant equipment and technology should be periodically checked to ensure that it is up-to-date and operable and to replace whatever is not.

The local emergency manager and other designees should be alert to any changes in roles, responsibilities, and rules in other local governments with which his or her government will have to interact in an emergency. That information has to be documented, shared, and reviewed with other responsible local government personnel.

Planning for Generally Predictable Events with Generally Predictable Consequences

Local government officials in Ohio should not spend any time planning for what to do when the next hurricane brings a storm surge through Shelby: it won't happen. However, if their city is on the Ohio River, they might want to think about what to do after the next flood. Assuming they already have a process plan in place, they need to focus on what happens when communities along the river or its tributaries become flooded. Most floods in a given locale seem to repeat themselves with at least some degree of similarity. It should not be hard to create realistic scenarios of what is likely to happen, when it is likely to happen, and what demands will be placed on local government.

It is not an accident that the state of Florida and its local governments are taking the lead in pre-event disaster recovery planning for hurricanes. Hurricanes are common events along the Gulf and southern Atlantic coasts, and they occur in Florida frequently. Thus, it makes considerable sense for communities in those areas to make plans for what to do when a hurricane threatens or actually strikes. And there is a substantial database from which to draw inferences about what is likely to happen should a hurricane strike and what will be needed in its immediate aftermath. Some hospital executives that we interviewed after Katrina described their organizations' efforts to train people in New Orleans and along the Mississippi coast according to what they had learned from hurricanes in Florida. Those hospitals, while flooded, recovered and reopened in a timely fashion. They learned from others' mistakes and successes. So, too, can municipalities learn.

Various parts of the country are subject to various hazards. Some places are seismically active and can expect moderate earthquakes periodically. Others have snoozing or snorting volcanoes nearby. Still others can expect a gray-green sky to spawn a tornado at almost any time in the spring and summer. Flash floods, storm surges, droughts, large industrial accidents, riots, epidemics, and terrorist attacks threaten numerous localities. The more likely it is that a particular event will occur in a community, the more attention that local officials should give to planning what should be done during the event and in the immediate aftermath.

Planning for what to do in the immediate aftermath of an extreme event is not the exclusive province of local government officials, though. If the community sewage treatment facility is near where the river floods, officials there need to be thinking in advance how to protect the facility, move it to safer ground, or operate when the floodwaters overtop the clarifiers or when the electricity needed to run the pumps and the aerators is inoperable. Offering training and involving such organizations in community-wide planning can jump-start recovery after an extreme event and strengthen ties between key community players and opinion leaders.

Planning for the most likely contingencies is critically important. We do not, however, encourage local governments to try to plan for all contingencies in advance. It can't be done. We can certainly anticipate the immediate and immediately following consequences of various kinds of events, but we cannot anticipate with any degree of reliability the nature, extent, and combination and permutations of the cascading consequences for the community system. Nevertheless, to the extent that we can speculate on what some of those consequences might be and which might be among the worst, we can appropriately take steps in advance to mitigate against losses in those critical components.

WHAT WE WOULD DO BEFORE THE NEXT EVENT

- *Plan.* We would work with a wide range of community stakeholders to develop comprehensive plans for the community's future. Such plans would include dramatic changes to the community that could be implemented in the wake of a large-scale disaster. We would use these plans to guide our annual planning and budgeting.

- *Diversify.* We would work hard to diversify the local economy, both by type of business and by business location. Too much dependency on one industry or too much density in one location seems to increase the likelihood of community collapse. The driving principle is that complex systems need slack in order to self-organize. Slack means lots of industry options, in lots of locations, requiring lots of different skills and expertise.

- *Network.* We would work with others to ensure that we will be able to restore utilities and local government services quickly. When disaster hits, we do not want to be scrambling to find out where to buy what and wondering what to do about restoring the most basic services and infrastructure.
- *Standardize.* We would standardize our equipment and technology so that we are not caught like New Orleans, with pump motors that no one else uses and that cannot be replaced easily. We do not want to find ourselves with a communication system with which we cannot communicate outside our own community.
- *Position Your Community.* We would make certain that we are in the National Flood Insurance Program (NFIP). Unless a community is part of the program, residents cannot obtain the government's subsidized flood insurance. The citizens of Lake Delton, Wisconsin, did not have flood insurance when the lake emptied catastrophically into the Wisconsin River because the city had dropped out of the NFIP seven years earlier.
- *Partner.* We would make sure that contracts are in place with suppliers and repair firms and that those contracts will hold up in the event of a large-scale disaster. This will help guarantee that we don't end up not being able to evacuate, for example, because all of the organizations have contracted with only one firm to evacuate our key personnel and clients, and that firm doesn't have sufficient assets to perform.
- *Start Today.* We would make building resiliency into our community part of every public investment decision. Each day, the community can be made a little more resistant to the adverse consequences of an extreme event. We can't take all the precautions against disaster damages, but we can take many of them incrementally. And we would begin today.

CONCLUDING WORDS

While the recovery challenge exists at individual, family, organizational, community, regional, and national levels, our primary concern in this book has been with community recovery. Community recovery means somewhat different things in different places. It is not just a matter of rebuilding structures and restoring services. It does not occur on a predefined timetable. To us, community recovery occurs when the community system develops long-term viability in the post-event milieu at a level that is roughly consistent with the expectations that the residents have developed over time after the event. And it is not guaranteed: not every community that attempts to recover is successful.

The recovery challenge is defined, in part, by the nature and extent of the consequences of the extreme event. Communities that do not suffer cascading consequences that damage the social, economic, and political community have a relatively easy job ahead of them. When there is damage to the fabric of the community, recovery requires much more than rebuilding and restoring the built environment. In those cases, local government and community leaders can lead, steer, facilitate, and work to bend the community "twig" so that it develops in a certain direction, but they cannot guarantee the outcomes of recovery efforts.

To a very considerable extent, the recovery challenge is shaped by the pre-event characteristics of the community, coupled with the goals and expectations of those in the community for the community. Those communities that were struggling for social, economic, or political viability before the event are likely to still be in trouble after it.

Finally, in our years of studying long-term recovery in communities across the country, we were greatly inspired by the level of commitment, the persistence, the intelligence, and the personal integrity of so many local officials and community leaders. Our hope is that compiling what we have learned from them in this book will help them and others in their efforts to recover and reclaim their communities.

ENDNOTES

1 Martin Wolk, "Are Disasters Really Good for the Economy?" MSNBC, September 9, 2005, http://www.msnbc.msn.com/id/9271060/ (accessed November 30, 2008).

2 J. Eugene Haas, Robert W. Kates, and Martyn J. Bowden, *Reconstruction following Disaster* (Cambridge: MIT Press, 1977).

3 Morris Freilich, "Toward an Operational Definition of Community," *Rural Sociology* 28, no. 2 (1963): 117–127, as cited in Robin Hamman, "Introduction to Virtual Communities Research and Cybersociology Magazine Issue Two," *Cybersociology Magazine,* November 20, 1997, http://www.cybersociology.com/files/ 2_1_hamman.html (accessed November 16, 2008).

4 Hamman, "Introduction to Virtual Communities Research."

5 George A. Hillery, "Definitions of Community: Areas of Agreement," *Rural Sociology* 20 (1955): 111–123, as cited in Hamman, "Introduction to Virtual Communities Research."

6 Haas, Kates, and Bowden, *Reconstruction following Disaster,* 10.

7 Jordan Lite, "Workers, ex-N.Y.ers from across the Nation Report WTC Ailments," *Daily News,* April 30, 2007, http://www.nydailynews.com/news/2007/04/30/ 2007-04-30_50state_ills-4.html (accessed November 30, 2008).

8 Ibid.

9 W. D. Iwan, H. Kanamori, W. K. Hudnut, et al., eds., "Summary Report on the Great Sumatra Earthquakes and Indian Ocean Tsunamis of 26 December 2004 and 28 March 2005," *Earthquake Spectra* 22, Special Issue 3 (June 2006).

10 Martin Savidge, "Oreck Brushes Off the Gulf Coast," *MSNBC.com,* February 23, 2007, http://www.msnbc.msn.com/id/17300566/ (accessed December 28, 2008).

11 Brad Kessie, "'Harsh Realities' Forced Oreck's Relocation Decision," WLOX-TV, December 13, 2006, http://www.wlox.com/Global/story.asp?s=5805473 (accessed November 30, 2008).

12 Susan Saulny, "New Orleans Hurt by Acute Rental Shortage," *New York Times,* December 3, 2007, http://www.nytimes.com/2007/12/03/us/nationalspecial/ 03renters.html (accessed November 30, 2008).

13 Lynne Jeter, "Coast Condo Market 'White Hot' Now, Growing Hotter," *Mississippi Business Journal,* August 28, 2006, http://www.allbusiness.com/north-america/ united-states-mississippi/4088844-1.html (accessed November 30, 2008).

14 Lousiana Speaks, "East Baton Rouge—Disaster Impact and Needs Assessment" (2006), http://www.louisianaspeaks-parishplans.org/IndParishHomepage_ BaselineNeedsAssessment.cfm?EntID=6 (accessed November 30, 2008).

15   Jacqueline Adams, "Tale of Two Cities: Biloxi and New Orleans," *CNN.com,* August 29, 2007, http://www.cnn.com/2007/US/08/29/katrina.twocities/ index.html#cnnSTCOther1 (accessed November 30, 2008).

16   Matt Rosenberg, "Sectors of the Economy," *About.com:Geography,* January 14, 2007, http://geography.about.com/od/urbaneconomicgeography/a/ sectorseconomy.htm (accessed December 5, 2008).

17   Allison Plyer, "March 2008 New Orleans Population Growth Slow in First Quarter 2008," Greater New Orleans Community Data Center, April 24, 2008, http://www.gnocdc.org/media/GNOCDCApr24-08.pdf (accessed December 5, 2008).

18   Adam Nossiter, "Whites Take a Majority on New Orleans's Council," *New York Times,* November 20, 2007, http://www.nytimes.com/2007/11/20/us/nationalspecial/ 20orleans.html?_r=1&ref=us&oref=slogin (accessed December 5, 2008).

19   Peter Whoriskey, "New Orleans Repeats Mistakes as It Rebuilds," *Washington Post,* January 4, 2007, http://www.washingtonpost.com/wp-dyn/content/ article/2007/01/03/AR2007010301593.html (accessed December 6, 2008).

20   FEMA describes the process by which a Presidential Disaster Declaration is made: "Once a disaster has occurred, and the State has declared a state of emergency, the State will evaluate the recovery capabilities of the State and local governments. If it is determined that the damage is beyond their recovery capability, the governor will normally send a request letter to the President, directed through the Regional Director of the appropriate FEMA region. The President then makes the decision whether or not to declare a major disaster or emergency." (See http://www.fema.gov/government/grant/pa/pr_declaration.shtm.)

21   Biloxi, Mississippi, *Comprehensive Annual Financial Report: City of Biloxi, Fiscal Year Ended September 30, 2006,* http://www.biloxi.ms.us/PDF/CAFR06.pdf (accessed December 13, 2008).

22   Ed Rappaport, "Preliminary Report: Hurricane Andrew, 16–28 August, 1992" (Miami, Fla.: National Centers for Environmental Prediction, National Hurricane Center, National Oceanic and Atmospheric Administration, December 10, 1993), http://www.nhc.noaa.gov/1992andrew.html (accessed December 13, 2008).

23   Joint Legislative Audit Bureau, *Local Government User Fees* (Madison, Wis.: April 2004), 3, http://www.legis.state.wi.us/LAB/reports/04-0UserFeesFull.pdf (accessed December 14, 2008).

24   Becky Gillette, "Business Interruption Insurance Can Save the Day—or Not," *Mississippi Business Journal,* July 17, 2006, http://www.allbusiness.com/ north-america/united-states-mississippi/4073148-1.html (accessed December 14, 2008).

25   Federal Emergency Management Agency (FEMA), "Community Disaster Loan Program" (November 2008), http://www.fema.gov/government/grant/fs_cdl.shtm (accessed December 14, 2008).

26   U.S. General Accounting Office, *Fire Management: Lessons Learned from the Cerro Grande (Los Alamos) Fire,* GAO/T-RCED-00-257 (Washington, D.C.: July 20, 2000), 34, http://www.gao.gov/new.items/rc00257t.pdf (accessed December 14, 2008).

27   FEMA, "Northern New Mexico Communities Remember Cerro Grande Fire," May 5, 2005, http://www.fema.gov/news/newsrelease.fema?id=17417 (accessed December 14, 2008).

28 FEMA, "FEMA Director Signs Final Cerro Grande Fire Compensation Regulations," press release ,March 15, 2001, http://www.fema.gov/news/newsrelease. fema?id=10047 (accessed December 15, 2008).

29 Havidán Rodríguez and Russell Dynes, "Finding and Framing Katrina: The Social Construction of Disaster," Social Science Research Council, June 11, 2006, http://understandingkatrina.ssrc.org/Dynes_Rodriguez/ (accessed December 29, 2008).

30 Frank Dexter Brown, "The Destruction of Princeville," *Seeingblack.com,* April 9, 2001, http://www.seeingblack.com/x040901/princeville.shtml (accessed December 28, 2008).

31 Whoriskey, "New Orleans Repeats Mistakes."

32 FEMA, *Disaster Assistance: A Guide to Recovery Programs,* FEMA 229(4) (Washington, D.C.: FEMA, September 2005), http://www.fema.gov/pdf/rebuild/ltrc/ recoveryprograms229.pdf (accessed December 24, 2008).

33 *Tarboro* (North Carolina) *Daily Southerner,* November 12, 1999, Flood–Special Edition, 9.

34 Small Business Administration, Disaster Loan Program, *Code of Federal Regulations,* title 13, vol.1, part 123.1, rev. January 1, 1999 (accessed December 27, 2008).

35 Ibid., part 123.6.

36 Sandro Galea, Arijit Nandi, and David Vlahov, "The Epidemiology of Post-Traumatic Stress Disorder after Disasters," *Epidemiologic Reviews* 27, no. 1 (2005): 78–91, http:// epirev.oxfordjournals.org/cgi/content/full/27/1/78 (accessed December 28, 2008).

37 Macon County Chamber of Commerce, "Montezuma," Macon County Web site (2008), http://www.maconcountyga.org/Histmtz.html (accessed December 27, 2008).

38 Ibid.

39 City of St. Peter, Minn., Comprehensive Plan (1995).

40 City of St. Peter, Minn., Comprehensive Plan (2004), 7.

41 Robert C. Nesbit, *Wisconsin: A History* (Madison: University of Wisconsin Press, 1973).

42 *Holistic Disaster Recovery: Ideas for Building Local Sustainability after a Natural Disaster* (Fairfax, Va: Public Entity Risk Institute, 2006).

43 Peter Wagner and Susan Edwards, "New Orleans by the Numbers," *Dollars & Sense* (March/April 2006), http://www.dollarsandsense.org/archives/2006/ 0306wagneredwards.html (accessed December 28, 2008).

44 W. Ross Ashby, "Variety, Constraint, and the Law of Requisite Variety," in *Modern Systems Research for the Behavioral Scientist,* ed. Walter Buckley (Chicago: Aldine Publishing Co., 1968).

45 Daniel J. Alesch and William J. Petak, *The Politics and Economics of Earthquake Hazard Mitigation* (Boulder: Institute of Behavioral Science, University of Colorado, 1986).

46 Lucy A. Arendt, Daniel J. Alesch, and William J. Petak, *Hazard Mitigation Investment Decision Making: Organizational Response to Legislative Mandate,* MCEER-07-002 (Buffalo, N.Y.: Multidisciplinary Center for Earthquake Engineering Research, 2007).

47 Daniel J. Alesch et al., *Organizations at Risk: What Happens When Small Businesses and Not-For-Profits Encounter Natural Disasters* (Fairfax, Va.: Public Entity Risk Institute, 2001).